Science and Socio-Religious Revolution in India

Scholars have long noticed a discrepancy in the way non-Western and Western peoples conceptualize the scientific and religious worlds. Non-Western traditions and communities, such as of India, are better positioned to provide an alternative to the Western dualistic thinking of separating science and religion. The Himalayan Environmental Studies and Conservation Organization (HESCO) was founded by Dr. Anil Joshi in the 1970s as a new movement looking at the economic and development needs of rural villages in the Indian Himalayas, and encouraging them to use local resources in order to open up new avenues to self-reliance.

This thoroughly-revised text argues that the concept of dharma, the law that supports the regulatory order of the universe in Indian culture, can be applied as an overarching term for HESCO's socio-economic work. This book presents the social-environmental work in contemporary India by Dr. Anil Joshi in the Himalayas and by Baba Seechewal in Punjab, combining the ideas of traditional and scientific ecological knowledge systems. Based on these two examples, the book presents the holistic model transcending the dichotomies of nature vs. culture and science vs. religion, especially as practiced and utilized in the non-Western society such as India.

Using the example of HESCO, the book highlights that the very categories of religion and science are problematic when applied to non-Western traditions, but that Western technologies can be radically transformed through integration with regional legacies to enable the flourishing of a multiplicity of knowledge-traditions and the societies that depend upon them. It will be of interest to students and scholars of South Asian Studies, Religion, Environmental Studies, Himalayan Studies, and Development Studies.

Pankaj Jain is Associate Professor in the Department of Philosophy and Religion at the University of North Texas, USA.

Routledge Studies in Asian Religion and Philosophy

For a full list of titles in this series, please visit www.routledge.com/series/rfbm

Science and Socio-Religious Revolution in India

Moving the Mountains

Pankaj Jain

Routledge
Taylor & Francis Group

LONDON AND NEW YORK

First published 2017
by Routledge

2 Park Square, Milton Park, Abingdon, Oxfordshire OX14 4RN
711 Third Avenue, New York, NY 10017

Routledge is an imprint of the Taylor & Francis Group, an informa business

First issued in paperback 2018

British Library Cataloguing-in-Publication Data
A catalogue record for this book is available from the British Library

Library of Congress Cataloging-in-Publication Data
Names: Jain, Pankaj, author.
Title: Science and socio-religious revolution in India : moving the mountains / Pankaj Jain.
Description: New York : Routledge, 2017. | Series: Routledge studies in Asian religion and philosophy ; 20 | Includes bibliographical references and index.
Identifiers: LCCN 2016041909 | ISBN 9781138023598 (hardback) | ISBN 9781315776316 (ebook)
Subjects: LCSH: Ecology—Religious aspects. | Religion and science—India. | Himalayan Environmental Studies & Conservation Organisation. | Mountain ecology—Himalaya Mountains Region.
Classification: LCC BL65.E36 J35 2017 | DDC 201/.77—dc23
LC record available at https://lccn.loc.gov/2016041909

ISBN: 978-1-138-02359-8 (hbk)
ISBN: 978-0-367-02604-2 (pbk)

Typeset in Times New Roman
by Apex CoVantage, LLC

Contents

Figures

Acknowledgements

Soon after I joined the University of North Texas in 2010, the department of philosophy and religion organized a conference for the Society for Philosophy and Technology. Dr. Anil Joshi, founder-director of HESCO (Himalayan Environmental Studies and Conservation Organization), came from India to present his work for this conference. Impressed by his scientific and social work in Indian Himalayas, I applied and received the Fulbright Fellowship that enabled me to travel and stay in Uttarakhand for few months.

This book is my humble tribute to the natural and cultural resources of that Devbhumi, divine land of gods and goddesses. I will forever remain grateful to Dr. Joshi and his colleagues, Dr. Rakesh Kumar, Dr. Kiran Negi, Dr. Sudha Sharma, Dr. Himani Purohit, Vinod Khati, Manmohan Burky, Dwarika Semwal, Anju Shangari, Pankaj Kumar, Jagdamba Maithani, Jagat Ramoul, Prem Kandwal, Shersingh Rawat, and many others, who opened their hearts and homes for me. Many of them travelled widely with me to several towns and villages of Uttarakhand where we met farmers and their families who continue to chart their destinies even as climate change increasingly threatens their daily lives. During that trip, I also met Baba Seechewal who continues to work tirelessly in Punjab to revive and restore water resources there. I dedicate one chapter to celebrate efforts of his Sikh community as well. I am indebted to Professor Purushottama Bilimoria for writing the foreword for this book. All the interviews and photographs were taken by me in 2012 in India. Some portions of the book have earlier appeared on the websites of both the organizations HESCO.in and NirmalKuteya.com as also acknowledged on these websites.

Last but not the least, I am also indebted to my wife Sonia and our two boys Atharv and Anav who continue to allow me to go on such long India discovery trips so that together we can continue to learn such great socio-ecological work being done amidst challenging times.

October 2nd, 2016 Pankaj Jain
(The Birthday of Mahatma Gandhi)
University of North Texas USA

Foreword

Is the war between science and religion a Western invention?

It is my great honour and pleasure to be able to pen a Foreword to this wonderful work – a sequel rather – by Dr. Pankaj Jain on *Dharma and Science*. To fill the missing "dharmic" gap in contemporary discourse of "religion and ecology," in his first book (Jain 2011) Dr. Jain had constructed a discourse based on *dharma*, both as a textual/historical and contextual/ contemporary phenomenon of Indic communities as well as an academic ethno-social scientific category in the study and interpretation of Indic cultures and their ecologies (see Chapter 2). In the present book-length study, the author in a similar vein sets out to fill the gap extant in contemporary discourses on "religion and science" by setting out a discourse based on *dharma* in intersectional dialogue with science as *science* is understood both in modern (post-modern) times, but also in the classical Indic context (from the Vedas onwards).

In this Foreword I wish to indulge the readers of this fine and timely work to the related discussion of a number of issues in the broader context of the tussle between science and religion that has become something of a mainstay of academic discourse in a number of disciplines, ranging from theology, religion, media, and peripherally to the sciences themselves. Equally significantly, it is worth exploring the question whether the scenario described as that of an ongoing "war" between science and religion is one entirely of Western invention – that is to say, something that is more true of the disciplines just mentioned as practiced in the West and in the modern academy – than it is in the Indian context and from the "Indic" perspective. Let me give some background to the quandary I am beleaguered with before suggesting a possible response to the question raised.

The New Atheism Movement in the West

A Mahābhārata-type of textual war has been raging for some time now, let us say since the early centuries of the European Enlightenment, between science (the sciences) and religion (theology and traditional ortho-doxatic systems). This war has many protagonists and antagonists, and its fangs have spilled over onto the Indian subcontinent as well, albeit with some differences. Both sides of the camp as it were engage in quasi-philosophical and other intellectual disputations to either drive the wedge to the death of one side, or falteringly attempt a rapprochement between the seemingly incommensurable claimants to the truth.

By far, the most vociferous voice today is broadcast broadly from the so-called Four Horsemen and their aligned New Atheist gangsters who echo the Enlightenment project of subjecting all truth-claims of religion-cum-theology based on faith and revelation to close scrutiny and testing them against knowledge delivered to us by science, based on observations, mathematics and rational deliberations. In this process the Enlightenment's secularizing edict of the absolute separation of Church and State could be ensured, and assuredly science sees itself as firmly vested in the "disenchantment of the world."[1]

The main advocates of the New Atheism Church are Harris, Dennett, Dawkins, and (the late) Hitchens,[2] – (no woman or non-whites are part of the clique); they are self-proclaimed agents not of God but of the temple of science. For their part, Dawkins and Dennett resort to evolutionary biology to explain the origin of the world and hence the improbability of a necessarily self-existent transcendent Being we call (as) God. To Dawkins, the universe appears to have emerged by way of "natural selection, the blind, inconspicuous automatic process" with "no purpose in mind" and no causal agency either.[3] Dennett goes beyond Charles Darwin in claiming that the algorithmic process of natural selection is alone sufficient to account for all design, without any need to appeal to intelligence, purpose or intentional contrivance for which theists invoke, literally, the Mind of God (not just a metaphoric metonym as in Paul Davis's cosmology – though Davis confesses that the "fine-tuning" and steady pace of entropy appear to presage a mathematically smart "intelligence" that remains inexplicable, or the universe would have been otherwise if not collapsed long ago[4]). Nature operates by its own finely tuned "cunning engineering" – its own "sky-hooks" and "cranes" – i.e. chance variations in genotypes that give special selectable advantage.

Even so, as some have critically argued, natural selection is only a theory of elimination of the weak and survival of the fittest; Darwin made no claims that one genotype necessarily without any purpose, or "blindly," mutates into another: i.e. it only reproduces itself with some variation and is never the *source* of the advantage enjoyed by the surviving genotype. Dennett surreptitiously sneaks in memes ("idea," borrowed from Dawkins) as some kind of built-in intentional function, which renders it *not*-blind at all. We have indeed observed adaptation in Darwinian finches (and politicians), but we have yet to truly observe the evolution of one species into another (even Man into Trumpian Übermensch, Hanuman to Ram, or Larry into *Dasein*), and that could well take some 600 million years. Hence we can only conjecture from fossils and life around us, and despite the evidence manifest in homologous and analogous structures, we can conclude on only some version of what Stephen Jay Gould called "local mutation mostly by gratuitous natural selection." No record of a new species having evolved is to be found in archives of known human history.

In any case, having amassed such water-tight "evidence" by a methodology that allegedly renders them unassailable, the New Atheists go further and claim that religion in the modern "Secular Age" (a slight parody on Charles Taylor's use of this trope) is not only antithetical, but an anathema to science, since it is obsolete and indeed detrimental to the evolutionary development of the human species, precisely because of the unfalsifiable, absolutist and eschatological nature of religious claims, which in turn have a retrogressive impact on the growth of civilization. Moreover, the superstitious and exaggeratedly unfounded beliefs, along with their archaic magical rituals, a mystical cosmology rife with non-existent beings and fabricated teleology, and so forth, only serves the patriarchal institutions with their patristic hierarchies of power to suppress the common person, including the gendered and under-class or caste, from a certain literacy and technological sophistication to find a niche in and contribute to the developed and developing social conditions. Religion, according to this view, is an aberrant evolutionary left-over seething in the bio-physiology of certain antiquated non-erectus *home sapiens* that breeds violence, nationalistic zealousness, jingioism, and fundamentalism to boot – as we witness these in the terrorism-torn world today; in short, religion is the self-lubricating axle of evil (ibid). This cynical view finds its echoes in the criticisms of Indian observers too when we get in the subcontinent. Thus, there is the critique from the likes of Meera Nanda and other sceptics who believe that there *ought* to be an outright war declared between the secular ethos and the idiosyncratic Indian claims of scientific discoveries in the texts and practices of religion going back allegedly to the ancient times, for, according to the skeptical view, the claims on the part of the latter are simply preposterous and unassailable.

The quasi-secular science and religion-friendly Indian turn

I have transposed in my 2011 *JICPR* paper precisely this cynicism and generally the contours of the modern Western embattled encounter of science and religion to the contemporary Indian context and recounted some of the voices on the sorts of questions and challenges that have been thrown up, beginning with late 19th- and early 20th-century encounters with the European sciences.[5] From Syed Ahmad Khan, Gandhi's verdict on technology, the scientization of meditation under Maharishi Mahesh Yogi (and now yoga elsewhere also), to the live-narratives of a handful of great scientists, notably, C. V. Raman, Chandrasekhar, J. C. Bose, Meghnad Saha, Vikram Sarabai, Ramanujan, Govindaraja, Abdul Kalam, the Sri Sathya Sai Superspeciality Hospitals, and with much borrowings from the eminent historian of Indian sciences, B. V. Subbarayyapa, I have demonstrated a totally different kind of response to the questions and war-cry that have bewitched the West. There is a new playfield evolving in the broadly Hindu (and to some extent Islamic and Sikh) contexts where emphasis is less on a *creator-visioned* universe that seems pitted against the *naturalized-evolutionist* view of the world, and more on a naturalized view of *religion* and a less *materialistic* vision of the universe (the vectors of the natural-supernatural are almost reversed). A bit like with the Mīmāṃsaka and Jain cosmologies, wherein their respective nontheism or agnosticism is not seen as a threat at all to orthodoxy but is rather welcomed as an alternative to all kinds of absolutism, providing there is room for meaning and spirit, and even for ritual in the pragmatics of day-to-day living (astrology and ecology in lieu of the tougher science of astronomy and natural sciences that might just do for after-hours *gṛhastāśrama* or the householder's obligatory callings).

Arguably it was the British colonial administration that introduced science into educational curriculum in the subcontinent, along with the use of science and burgeoning European technologies to produce surveys, construct irrigation canals, erect dams, collect census data, and build railways and later communication systems (printing press, moss-code, radio, telephone, royal mail, etc.). There was clearly a highly charged political agenda in their mission. Nevertheless, Indian scientists responded to the wave of nationalism that manifested itself in many ways, one of which was an inclination to design research so as to highlight peculiarities of the indigenous environment. Thus nativist voices emerged within the field of science and aligned epistemologies to disrupt colonial production of the modernist epistemology. For example, in the late 1880s Madhoo Singh II built the Jeypore (Jaipur) Museum in the "Pink City" (India's architectural modernism) to house *Razmnāma* (Book of War), an abridged Persian rendering of the *Mahābhārata* commissioned by Emperor Akbar in the 1580s and acquired by the Kacchhwala rulers of Jaipur from the Mughal Indo-Persian collections. And then there were the Tagores struggling to respond to the challenges of Darwinian evolutionary theory and the dogma of creationism.

The mainstreaming initiative, however, in the halls of Indian science came in the later half of the 19th century from a Calcutta-based MD turned homeopath, Dr. Mahendra Lal Sircar, and resulted in the establishment of the Indian Association for the Cultivation of Science in 1876.[6] Educated and elite Indians took home other benign messages from their exposure to the achievements of sciences in the West, and came also to know of (1) the contributions of Arabic-Islamic scientific discoveries and recoveries (from Greek sources) on Western science and the industrializing societies, and (2) emerging textual evidence (via German Indologists mostly) of the existence of possible comparable science in ancient India, as also in China. Indian scientists trained under the British system – such as Syed Ahmad Khan, B. N. Seal, J. C. Bose, Meghnad Shah, Srinivasamurthi – were aware of the hostility of Western science toward India and the ruse of science under colonialism (to undermine the confidence and traditional wisdom of the natives, anywhere – think of the famous or infamous Macaulay Minutes of 1835). Most believed on the contrary that traditional religion and science of the West could be reconciled without any losses to either. Syed Ahmad Khan, while a practicing quasi-mystic, and following the Islamic modernist, Jamaluddin Afghani, looked favorably to science to revive Muslim culture and literacy in the belief that the subcontinent could benefit from England's technological wealth.[7] But most Indian scientists throughout the colonial period lived an almost schizoid existence: scientists by day and theologians by night, as it were. Still, they plodded on, looking for a healthy and happy syncreticism, in the fashion of *miri-piri* (secularism/spirituality) formulated by Sikh Guru Arjan's son, Guru Hargobind, back in 1606.

So, for example, Srinivasamurthi, a Sanskrit scholar who in 1923 was dreaming of releasing atomic energy, turned his gaze toward indigenous medical systems, especially Āyurveda, and made a Galen-like stride in incorporating Hindu Caraka-Saṃhitā and Muslim medicinal practices into the modern paradigm of health management, informed by Buddhist non-dualist philosophical principles. He formulated a new paradigm of chemistry that was scathing of the germ theory so central to Western medicine. He changed Indian medical science forever. Meanwhile, Prafulla Chandra Ray (1861–1944), known as the "father of modern chemistry" in India, was building on his work on the compound mercurous nitrite.

Botanist and physicist Jagadish Chandra Bose (1858–1937) went on to find evolutionary scientific basis for the Indian-wisdom belief in the vital sentience of plant life, and he succeeded in proving his thesis of vitalism (*prāṇic energy*) which gained popular scientific attention, even though he was ridiculed by the British Academy of Science. But what is not that well-known about Bose is that he preceded Guglielmo Marconi in successfully transmitting electromagnetic waves from one place to another without the use of wires, or solid conducting agents; the only difference is that while Bose used the confines of his small home laboratory, Marconi succeeded in transmitting the electromagnetic waves across the Atlantic, making possible the invention of long-distance Morse-code-telegram, telephonic communications, facsimile transmission, and eventually the internet and your pocket cell-phone. But Marconi acknowledged Bose's unprecedented discovery. Yet another bhadralok Bose, Dr. Girindrasekhar, made strides in psychoanalysis, combining Freud's insights into the unconscious with Indian theories on yoga, the subtle-conscious body, and traditional psycho-spiritual practices of meditation and inner expansion. Bose communicated with Freud on his discoveries and likely mentioned terms such as Nirvana.[8] He founded the Indian Psychoanalytic Society in 1922 in Calcutta, along with various journals, as well as the Indian Psychoanalytic Institute in 1932 (for training analysts), and he established the first such OPD in Asia, all alive to this day.[9] Meghnad Saha (1893–1956), a grocer's son, hence no bhadralok himself, made a remarkable contribution to astrophysics with his famous "ionization formula," still widely used to this day (also founded the journal *Science and Culture*). On the other end of the subcontinent, in the Punjab, Sikh scientists, such as Puran Singh, were making their own strides, and achieved some remarkable breakthroughs in various fields, not least in agriculture and forestry, but also in the atomic field, from which pedigree we have the present-day leading nuclear scientist, Piara Singh Gill. And remember that low pay, inadequate resources and bureaucratic obduracy and thoughtlessness that undermined even individuals of distinction, the hurdles were numerous for any aspiring Indian scientist, and yet they carried out their duties, their ambitious pursuits, and published a huge number of papers on their respective discoveries as well as speculative forays – more than the Americas and Oceania of those days.

The Scientific Temper

The Nehruvian penchant for the "temper of science" – standing at once for the trope of "secular"– was written into the Indian Constitution in 1950. As Amrita Shah notes,[10]

> [T]he idea [that science was indeed the new disciplinary savior] was further perpetuated in the post-Independence era and enshrined in the "temples" of modern India: in power plants, hydroelectric dams, steel factories, institutes of technology and national laboratories. Jawaharlal Nehru, a powerful votary of science's transformative capabilities, urged scientists to think not only of the pursuit of truth but of bettering the lot of India's people.

In the rest of this Foreword I wish to dwell on the post-colonial period for its delivery of the good, bad, and the ugly to the new nation-state, and here the trenchant deconstructions from two Indian feminist critics will be visited, namely, Meera Nanda and Abha Sur.

Take the case of the brilliant mathematician, Srinivasan Ramanujan (1887–1920): he practiced a neat and non-dualistic science that has been the forté of Indian thought since at least the 8th century AD. One school of thought believes that his science relied as much on mysticism, metaphysics, and astrology as it did on the abstract ideas of mathematics and geometry to an extent. The intertwining of mathematics and metaphysics was a unique phenomenon that indeed troubled his academic host in Cambridge, for he claimed to have arrived at some of the unprecedented solutions in his dreams through the help of the goddess of learning (Sarasvatī). But in effect these were reworkings of solutions already known to traditional mathematics that were used for charting the paths of stars and planets in the sciences of astronomy-cum-astrology. Others, however, believe that Ramanujan's achievements were peculiarly of a rare solitary genius with no influence from the tradition. This issue is a subject of much debate among Indian critics and historians of science.

Then there was the formidable C. V. Raman, who discovered "scattering" – known also as "the Raman Effect," meaning when light traverses a transparent material, the deflected bits change in wavelength. Raman headed the prestigious Indian Institute of Science (IISc) in Bangalore, was elected a Fellow of the Royal Society in 1924, knighted in 1929, and awarded the Nobel Prize in Physics in 1930. In his lifetime, Raman would train more than 100 physicists, including eminent scientists like Homi Jehangir Bhabha (1909–1966) [no relation to the close living namesake] and Vikram Sarabhai (1919–1971). Both came to IISc from Cambridge during World War II and were not able to return; Sarabai came to observe cosmic rays in tropical latitudes, while Bhabha distinguished himself with publications on positron physics and the cascade theory of cosmic showers. In the 1940s Raman worked with Max Born on crystal dynamics at the IISc; but Born could not fathom the aesthetic excesses and non-Euclidian geometrical structures that Raman had charted in his spectroscopic experiments on diamonds. The two scientists of vastly different temperament fell out in the end; Born charged Raman of speculative dabbling and unnecessary mystification and tried to refute Raman's findings. Only now with advanced C-CAD and digital scanning processes have scientists begun to appreciate some of Raman's discoveries and his likely use of traditional intuitive or aesthetic critique of Western science. But there are strong views among Indian scientists and critics that Raman was something of an "authoritarian Brahmin" when it came to pure science, and his temperament hindered rather than helped progress of science in India of the 1940s–50s.

Indeed, Raman's autocratic hubris created something of a hullaballoo in his busy laboratory, and this impacted on the fate of three women graduate students at IISc, none of whom got awarded the PhD they came to work toward under Ramana. They were Anna Mani (never married and took up career in meteorology), Lalitha Doraiswamy (gave up her aspirations in science and married the eminent physicist Chandrasekhar, Raman's nephew, and moved with him to Chicago), and K. Sunanda Bai. Now the case of Sunanda Bai is an interesting one as it is painful also: she had done pioneering work in recording and analyzing the composite nature of the scattered spectrum of liquids; alas, late 1944, Bai took her life days before she was to leave for Stockholm to take up a fellowship she had been offered at the University there in recognition of her excellent work in the field, despite Raman. In her incisive study of the "Raman Epic," Abha Sur narrates how the three women were relegated to the lower levels of laboratory practice. Raman enforced a strict separation of sexes, so they worked alone, snatching a few hours of sleep under tables in the lab during overnight experiments. Mani put in long hours recording and analyzing fluorescence, absorption, and Raman spectra of 32-odd diamonds. Bai did pioneering work in recording and analyzing the composite nature of the scattered spectrum of liquids. Neither was awarded a doctoral degree – apparently on technical grounds.[11]

Nevertheless, a major research institute is named after – as some call him – the "Giant of Modern Indian Science," to whit the Raman Institute, which is situated next to IISc, in Bangalore, where two or more generations of eminent Indian scientists have been trained, and who have made notable strides in many branches of science.[12] One such advance has been made in astrophysics, notably through the work and teaching of the Nobel Laureate in Astrophysics, S. Chandrasekhar, who worked in Chicago since 1935 and influenced Oppenheimer (to whom the famous epiphanous quote from the *Bhagavadgītā* about the splendor of the thousand suns is attributed, when in fact it was Chandrasekhar who first drew attention to this analogue). Chandrasekhar promoted Asian research scientists, two of whom jointly won the Noble Prize in the field in advance of him (a sign rather of humility and modesty of a Guru even in the tradition of science!). Moving on.

Saffron Nationalism – appropriation and complicity of science á-la Meera Nanda

Meera Nanda turns her disenchanted secularist sabré on the complicitous alignments, in the name of Hindu tolerance and eclecticism, but more sinisterly, of science with waves of extreme Hindu – "blood-and-soil" Āryan – nationalism, as manifest in recent decades more stridently in the Hindutva ideology of the Sangh Parivar a.k.a Saffron Brigade. In her monographs *Prophets Facing Backward, Postmodernism, Science and Hindu Nationalism; Breaking the Spell of Dharma and other essays: A Case for Indian*

Enlightenment; and *The God Market: How Globalization is Making Indian More Hindu, Postmodernism and Religious Fundamentalism: A Scientific Rebuttal to Hindu Science*, capped up by a powerful essay "Making Science Sacred" (included in the just-out *Science in Saffron Skeptical Essays on History of Science*), Nanda seeks to demonstrate a pernicious collusion of home-grown Indian science, revisionary modernism, classical rationalism (mostly in neoliberal economics or what's called "liberalisation," thinking of Jagdish Bhagavati in his Columbia ivory-chair, and the baton now passed onto Narendra Modi, the recently elected Hindutva-RSS-backed Prime Minister of India), re-enchantment of nature (in the environmentalist movement), and even postmodernist, feminist, and postcolonial-subaltern scholars, with reactionary Hindutva, that has turned religion into an "atavistic battleground for political manipulation and power plays."[13] Nanda calls this movement "the emerging state-temple-corporate complex" and claims that it is "replacing the more secular public institutions of the Nehruvian era." It combines traditional pseudoscience, in particular the superstitions of Vedic science – the claim that it triggered the modern discovery of quantum mechanics – and soliciting government support for including astrology in educational curriculum, while advocating a radical revision of text-books, histories of Indian "civilization" used in schools. These and the overblown merits of Āyurveda are just some examples of the pursuit of the "scientific-ity" of neo-Hinduism – with exclusionary Hindu chauvinism, while ignoring the menace of the caste system and barbaric social practices associated with a plethora of religious praxis, that exclude the majority from a share of the so-tailored or the more conventional forms of scientific education (ibid). There isn't even a decisive stance taken by Indian scientists and secular intellectuals (she has scholars like Nandy in mind) against Creationism, albeit of Purāṇic and popular Hindu theistic variations. They don't need to, for we should note that it is only with middle-and-Navya Nyāya (circa 7th to 16th century), and to a lesser extent with the 12th-century Viśiṣṭādvaita Vedānta savant Rāmānuja, that we first begin to see decisive cosmo-teleological and transcendental arguments of any sophistication for belief in the temporal existence of an only just "creationist-type" Īśvara or God;[14] it was left to the Mīmāṃsakas and Śaṅkara to refute sṛṣṭī of any kind – be it creation -ex nihilo, - ex machina (the Demiurge/ Potter model), emanationism, emergentism, immaculate or coitus conception, or some version of evolutionism as extracted from Sāṃkhya ontology and developed via Hegel by Sri Aurobindo into evolutionary spiritualism.[15] Creationism would not sit comfortably either with pantheism or panentheism,[16] much less with polytheism, and Hindu "theisms" are mostly of these varieties; although, it is also true that one cannot help noticing how the Western import of the lexically vague "God" is increasingly incorporated into Hindu popular deistic quasi-Semiticized doxology and popular religious as well as literary cultures that border on monotheism ("Ram is our absolute God"). Still, "creationism" as a dogma is not a universal creedal proclivity of Indian religions, and therefore the String-and Plasma-issuing Big Bang cosmology and the theory of evolution is not seen as a threat to their cosmogenic/species-origins forays (mythological or philosophically informed); neither should they nor their scientific counterpart rush to an evangelically feisty defense of creationism. Even Indian Christians/and/Muslims fail to see why there is so much sermonic ink spilled over this very local-Western fetish.

Liberalization (economic reform in consonance with "free market economy") and globalization are intimately connected with, if not the propellers of, on the one hand, the Green Revolution, the rise of the middle class, software development, and outsourcing industrial successes. To this movement we may attribute the meteoric rise of "Nova Tech" cities such as Hyderabad and Bengaluru. On the other hand is the Right-ward shift both in domestic politics and foreign policy that aligns itself strategically closer to the U.S. (seeking exemption on nuclear capabilities productions and a greater share in the "free trade market"), and so on.[17] The long and short of Nanda's unrepentant – over a decade now – argument is that conservative religious ideology goes hand-in-Ram's-*baan* with a muted scientific indolence and aggressive religious nationalism; this trend of course is growing in the U.S. as well; the majority of NR Indian scientists; Hindu, Jain, and Christian doctors; techies; and entrepreneurs in the U.S. tend to support some version of Creationism (and pro-life choices) and have of late crossed over to support Republican's, not President Obama's, neoliberal agenda with an eye to the benefits of the blinding globalization afoot.

I should like to conclude with these remarks. In January 2016, I attended an international conference on "Science and Jain Philosophy" held at the IIT in Mumbai. I was intrigued by the high level of interactive research-based presentations that were made by a number of eminently qualified scientists, medical professionals, monks, academics, and other teachers, all of Jain background (barring a few who were Europeans or otherwise Hindus). The conference explored, among other topics, the implication of Jain ethical practices of minimalist interference in nature to the larger challenges looming today in the areas of climate change, global warming, ecology, and animal justice. A day-long symposium looked at the use of *preksha* meditation for rehabilitation of health and behavioral deficiencies, control of passions, and the tracking of subcortical cognitive processes that affect one's normative performances in everyday life. The Jains gathered there (and others cited) seem to have worked up thoughtful answers to every one of the challenges and innovative forays being made in the contemporary scientific and our modern world – even thinking about Jain philosophy of time vis-à-vis Hawkings' history of time, the contributions enabled via Jain cosmology to certain glaring lacunae in the "Big Bang" theory, to rethinking universal ethics in a multicultural society. Now here, as is obvious, there is no dwelling on the seemingly interminable "war" between science and religion; rather there is an espousal and endorsement of the possibility of finding common ground between science and religious perspectives with their attendant practices and even mutually enhancing, if not simply embellishing, the findings on both sides of the "veil of inquiry" – the tools, resources, and panacea (even if this trajectory might be stretching the vision a bit far) for human flourishing. Much work is being done in these areas, and yet more remains to be done in this direction; but these are promising gems and in no way is the dialogue hampered by the kind of skepticism that Meera Nanda continues to propagate in her proliferate work; her target no doubt is the "Hindu saffron brigade"[18] and their claims to being precursors to numerous modern-day scientific discoveries and theories, and so forth. But why would she ignore the large "Dharma" picture and not include interlocutors from Jain (and Buddhist) and indeed pre-Hindu Indus and Vedic civilizational cultures from the intercepted dialogue? The chapters that follow attempt in their own way to rectify the imbalance and asymmetry in the scholarly treatment of the erstwhile and present-day relationship between dharma and science.

Purushottama Bilimoria

Notes

1 "War & Peace Between Science and Religion: The Divine Arch after the Four Horsemen," *Journal of the Indian Council for Philosophical Research*, XXVIII (2): 3–30 (April–June 2011). I draw also on a paper presented at the American Academy of Religion in 2012: "All India Radio: *The War between Science and Religion in the subcontinent – a Western import?*" My 2007 paper is amply referenced in chapter two of this book (see also note 12).
2 See, e.g., Sam Harris, *The End of Faith* (2004); Richard Dawkins, *The Blind Watchmaker* (1996), *The God Delusion* (2006); Daniel C. Dennett, *Breaking the Spell: Religion as a Natural Phenomenon* (2006); and Christopher Hitchens, *God Is Not Great* (2007).
3 Dawkins, *The Blind Watchmaker*, 5–6; and he claims Darwin made it possible to be and intellectually fulfilled atheist, which is not true about Darwin as he simply subverted the relationship between metaphysics and evolution. I discuss this conundrum in more detail in my *JICPR* paper (see note 1).
4 Paul Davis, *The Mind of God*. Even if the arrangements were different by 10^{-2}, the laws of physics and everything we know about the universe, including ourselves, would have been different if not non-existent.
5 See note 1 above.
6 Rajesh Kochhar, "Cultivation of Science in the 19th Century Bengal," *Indian Journal of Physics*, 82(7): 1003–1082 (2008).

7 See S. Irfan Habib, "Reconciling Science With Islam in 19th Century India," *Contributions to Indian Sociology*, 34(1): 63–92 (February 2000).

8 Ashis Nandy, *The Savage Freud and Other Essays on Possible and Retrievable Shelves* (New Delhi: Oxford University Press, 1998).

9 See Warwick Anderson, Richard C. Keller, and Deborah Anderson (eds), *Unconscious Dominions, Psychoanalysis, Colonial Trauma, and Global Sovereignties* (Durham, NC: Duke University Press), chapter by Chrisitiane Hartnack, "Synergies of Freudian Theory with Bengali Hindu Thought in British India."

10 Review of Abha Sur, *Dispersed Radiance: Women, Caste and Modern Science in India* (New Delhi: Navrang Publishing, 2009) by Amrita Shah in *Caravan A Magazine of Politics and Culture*, 1 November 2011, and retrieved from http://www.thehindu.com/arts/books/article2548119.ece

11 Culled from her lecture at UC-Berkeley to AID Meeting (2012); her book, and the review by Amrita Shah thereof. See previous note. See note above for full bibliographical details.

12 There is a short history of Raman Institute with a biographical sketch of his scientific work by G. Venkataraman, "Spirit of a Giant – C.V. Ramana," in *Traditions of Science, Cross-cultural Perspective*, ed. P. Bilimoria and M. K. Sridhar (New Delhi: Munshiram Manoharlal, 2007).

13 Ralph Dumain, Review of Meera Nanda, *The Wrongs of the Religious Right: Reflections on Science, Secularism and Hindutva* (Gurgaon, Haryana: Three Essays Collectives, 2005), *Logos, A Journal of Modern Society and Culture*, 9(1) (2010); and http://www.hindu.com/br/2006/11/21/stories/2006112100521400.htm

14 See P. Bilimoria, "Nyaya and Navya-Nyaya," in *Brill Encyclopedia of Hinduism*, ed. Knut Jacobsen et al., Vol III, pp. 657–671 (Leiden: E J Brill); and "Toward an Indian Theodicy," in *A Companion to The Problem of Evil & Theodicy*, ed. Justin McBrayer and Daniel Howard-Snyder, pp. 306–317 (Oxford: Wiley-Blackwell).

15 See P. Bilimoria, "Hindu Doubts About God: Towards a Mimamsa Deconstruction," *Indian Philosophy A Collection of Readings*, pp. 87–106 (New York: Garland Publishing Inc.); and in-tandem the two sources in above note, and item in note 1 also.

16 P. Bilimoria with Ellen Stansell, "Suturing the Body Corporate (Divine and Human) in the Brahmanic Traditions," Special Issue on Panentheism and Panpsychism, *Sophia*, 49(2): 237–259.

17 See comments in his Review of Sur, by T. Jayaraman, "Circular Reasoning," in *Front Line* (India), January 1, 2010: 80.

18 See her most recent, *Science in Saffron Skeptical Essays on History of Science* (Gurgaon: Three Essays Collective, 2016).

1 Introduction

Quoting a few lines from my 2013 article "The Environmental Sustainability of Spirituality," at HuffingtonPost.com:

> As an immigrant, I arrived at New York's JFK Airport on November 19, 1996, and was immediately struck by the hundreds of cars all around the airport plying on various "spaghetti" flyovers and highways. Born in a small sleepy town of Pali in Rajasthan and having lived most of my life in small towns in India, I was ready for all cultural shocks, the very first being the environmental one. I asked my friend Ajay who had come to pick me up, "How exactly will all these cars be sustained once the fuel supply is over?" Ajay, having arrived just a couple of months before me also from India as a software engineer like myself, proudly declared "Oh! This is America! They can run their cars even on water, don't worry!" Such is the faith of many Indians, Americans, and others who rely on modern science and technological aids such as cars and cell phones, mostly invented in America. With the impending environmental crisis looming large over humankind, is this faith weakening in the second decade of the 21st century, almost 20 years after my first American encounter?

For my first book project (Jain 2011), I spent much of my time and energy away from all the scientific and technological interventions in the villages of Rajasthan and Gujarat. My focus was the three communities of Bishnois, Swadhyayis, and Bhils, all of which are taking care of their natural resources inspired by their spiritual traditions, largely devoid of any modern technological interventions and oblivious of any environmental concerns arising out of urban consumption and industrial pollution. I had presented these three communities as three assertions against modernity. I called their environmentalism *Dharmic Ecology*, different from other kinds of environmentalism because the former kind, as practiced by the Bishnois, Swadhyayis, and Bhils, was really an expression of their dharma rather than a reaction against any environmental crisis. In contrast, most other kinds of environmentalism are driven as a reaction against a very specific issue, such as the Chipko Movement, the Appiko Movement, Save Narmada Movement, and various anti-nuclear and anti-chemical movements. Almost immediately after the publication of my first book, I came across Dr. Anil Joshi (hereafter Dr. Joshi) who had come to my university (University of North Texas) to present at a conference on technology and philosophy in May 2011. I was intrigued by the various projects in which he was using science and technology to help sustain the Himalayan environment by his initiative called Himalayan Environmental Studies and Conservation Organization (HESCO). Fortunately, I was selected for the Fulbright-Nehru Environmental Leadership Fellowship for the summer 2012 in which I conducted a detailed study of HESCO and its various projects in Uttarakhand and Himachal Pradesh, the two Himalayan states of India where HESCO has been working since 1970s.

And on June 12, 2012, I found myself suddenly transplanted into the Himalayas: from the hot and dry climate of Texas (where I presently live) and Rajasthan (the state where I was born and had conducted my earlier research) to rivers, mountains, and forests of the Himalayas, from purely dharmic ecology to scientific environmental work of HESCO. There could not have been a more different theme for my new research project. But this roller coaster ride is not just across the diverse climatic conditions of India; the ride that I am proposing in this book is across several other valleys and hills, such as the seemingly antagonistic dichotomies of religion and science and the seemingly polar opposites of nature and culture. An overall theme that I would like to continue in this book also is that of applying Western categories of knowledge to the study of non-Western cultures such as India, including "religion," "ethics," "science," "environment," "forest," "agriculture," and so on. As I wandered from village to village, from river to river, from mountain to mountain, from farm to farm across the Himalayas in Garhwal and Kumaon in Uttarakhand (one of the newest Indian states carved out of UP in 2000) and Himachal Pradesh (an Indian state that emerged in 1971), the categories to bound and describe the Himalayan ecology continued to transcend. I was never sure where a village started or ended, where its farms were tilled, or where its forests were; almost all these categories overlapped and transcended. Similarly, the work of HESCO, which makes substantial use of the latest technologies such as nuclear isotopes, is deeply rooted in the dharmic philosophy of life and death, such as rebirth and deep reverence for ancient traditions such as *gharat*, small watermills sustained by the local energy technologies. Also, unlike the mainstream scientific impetus to grow more genetically modified (GM) crops, HESCO has been working with local farmers for decades to encourage growing local grains such as *mandwa* (millet). During my stay in India, I also came across interesting socio-environmental work done by a Sikh religious leader Baba Balbir Singh Seechewal in Punjab, which I present in chapter 5.

With my undergraduate studies in computer science and graduate studies in Indic religious traditions, in this book, I hope to transcend the boundaries of "science" and "religion" as they apply towards the Himalayan environment, as they appear in HESCO's work, and as knowledge categories in American academic discourse. As the Sanskrit poet Kalidasa had referred to Himalaya as a mountain spanning the East and the West, I hope this book will be a humble contribution in greater understanding between Eastern (i.e., Indic) and American cultural and academic categories and phenomena. Agreeing with Weightman and Pandey (1978), I suggest that dharma can be a better paradigm with its kaleidoscopic meanings of physical characteristic, duty, virtue, religion, ethic, virtue, nature, cosmic law (i.e., *Ritam* of the Rigveda which *Dharma* replaces as a synonym eventually), and sustaining force (thus relating to sustenance and sustainability).

2   Dharma and science

From the earliest Sanskrit texts Rigveda to later Sanskrit texts about epics and myths, to contemporary Indian languages spoken by more than one billion Indians in India and elsewhere in the world, from the most ancient Jain texts based on Mahavira's statements to Buddhist texts based on Buddha's statements to medieval and contemporary Sikh texts, if there is one word that has fascinated peoples of India, it is *dharma*. With the spread of Buddhism across much of Asia, the term has continued to be used in scholarly texts and colloquial usage for more than two thousand years. Yet even after inclusion of this term in Oxford and other English dictionaries, this word continues to elude Westerners in their discourses of India and the rest of Asia. Although the academic literature on themes such as "religion and ecology" and "religion and science" has been actively produced in last several decades, there is hardly any academic book or article on the themes of "dharma and ecology" or "dharma and science." Even with the multidimensional and multivalent interpretations of dharma, ranging from religion, ethic, virtue, physical characteristic, nature, cosmic law, and sustaining force, the category of "religion" with its different semantic continues to dominate the English discourse about Asia.

To fill this "dharmic" gap, in my first book (Jain 2011) I had attempted to set the discourse based on dharma, both as a textual/historic and contextual/contemporary phenomenon of Indic communities, as well as an academic ethno-social scientific category to study and interpret Indic cultures and their ecologies. In the present book, in a similar vein, my attempt is to interpret the social-environmental work by HESCO in the Himalayas and by Baba Seechewal in Punjab from the "dharmic" category rather than religious. In this chapter, I will glance through some of the scientific achievements of 19th and 20th centuries and the recent critiques of science and religion in the Indian context.

To get to the root of dharmic science, I looked up the Sanskrit term for "science" and found *vijñāna*. According to the Monier Williams dictionary, the term first occurs in one of the earliest Indic texts, *Atharva Veda*, and means the act of distinguishing or discerning, understanding, comprehending, recognizing, intelligence, and knowledge. In the *Uttamacharitrakathānak*, it means skill, proficiency, and art. In the *Sushruta Samhitā*, it means science and doctrine. In the *Mahābhārata*, it means worldly or profane knowledge and the faculty of discernment or of right judgment. In the *Bhāgavat Purāṇa*, it means the organ of knowledge. The ancient Indic linguist Pāṇini defined *vijñāna* as the understanding of (a particular meaning). The Buddhist text *Dharmasamgraha* defined it as consciousness or thought-faculty (one of the five constituent elements or *Skandhas*, also considered as one of the six elements or *Dhātus*, and as one of the twelve links of the chain of causation). My attempt in this book is to explore all these meanings of *vijñāna*, the Indic term for science, through my fieldwork with HESCO in the Indian Himalayas and with Baba Seechewal's initiatives in Punjab.

In recent Indian history, as cited in Bilimoria (2007), it was the British colonial administration that introduced science into educational curriculum in the subcontinent as a separate category from other humanities-based courses. In the 19th century onwards, other European technologies were also whole-heartedly adopted to produce surveys, construct irrigation canals, erect dams, collect census data, and build railways and communication systems (such as the printing press, Morse code, radio, and telephone). However, in the wake of nationalistic movements, Indian scientists developed an indigenous research agenda, disrupting the colonial production of the modernist epistemology (Turaga 2005). In 1876, a Kolkata-based medical doctor Mahendra Lal Sircar founded the Indian Association for the Cultivation of Science (Kochar 2008). Many Indians also started discovering the contributions of Arabic-Islamic scientists, as well as emerging textual evidences of the existence of sciences and technologies in ancient India. Indian scientists trained under the British system – such as Syed Ahmad Khan, Brojendra Nath Seal, Jagadish Chandra Bose, Meghnad Saha, and Srinivasa Murthi – were aware of the colonial agenda to undermine the traditional wisdom of India. However, instead of rejecting European science, they tried to reconcile it with traditional Indic science and culture. For example, in 1923, a Sanskrit scholar Srinivasamurthi focused on indigenous medical systems, especially Āyurveda, and incorporated Hindu Caraka and Muslim medicinal practices into the modern paradigm of health management. Being critical of Western germ theory, he formulated a new paradigm of chemistry and changed Indian medical science forever. In this example, we see early Indian resistance against the onslaught of Western sciences (Bilimoria 2007).

In my ethnographic observations with HESCO, I noticed that science and technology play a prominent role for the Himalayan communities, and HESCO's founder and most of his colleagues are all trained in botany and other sciences and technologies. This combination of science and technology for social work is similar to the enthusiastic acceptance of science by the major social leaders in late 19th- and 20th-century India as described previously (Dorman 2011). Unlike the popular perception of regarding traditional religious views as inherently antagonistic to science, Indic traditions rarely if ever had the dichotomy between the two. Moreover, the very categories of religion and science are problematic when applied to non-Western traditions. Balagangadhara (1994) has argued that the term *religion* is not appropriate to non-Christian societies such as India. Similarly, the term *science* has evolved and is applied differently in the Indian context than it is in the West. For instance, with the advent of postmodern interpretations of science after the discoveries of quantum physics, Indian scientists have become more spiritual, as shown by Gosling (2007). As Dorman (2011:614) poignantly argues,

> such an opposing trend seems to signify a different fundamental worldview at the core of Indian scientists that does not lead to the same conclusions made by those in the West. Therefore, instead of consistently applying assumed Western preconceptions of science and religion to non-Western instantiations of the dialog, we need to recognize that science and religion have their own unique characterizations in Indian thought. These characterizations must be considered in their own right if we are to progress the field.

Such critiques of *science* from a non-Western perspective were severely challenged by scholars such as Meera Nanda (2003). For instance, if non-Western scientific practices demonstrate the intertwined relationship of nature and culture (or science and religion), instead of divided along the familiar Western Cartesian dichotomy, it was because they had not yet modernized by adopting the latest sciences and technologies. The local scientific practices should be represented as ethno-science rather than Western science based on universal Enlightenment ideals (Prasad 2006).

The efforts of the theoreticians of alternative sciences, such as Ashis Nandy, have been to look for epistemological alternatives to modern science in order to search for non-Eurocentric ontological possibilities. In his book, Nandy (1995) analyzes the biographies of two Indian scientists, Jagdish Chandra Bose and Srinivasa Ramanujan, whose epistemological methods greatly differed from Western empirical methods. Similarly, Brown in his book (2012) cites philosopher Jitendra Nath Mohanty, Indologist Wilhelm Halbfass, and various Hindu thinkers such as Keshab Chandra Sen, Dayananda, Vivekananda, and Ranganathananda to show the general Hindu perspective that science and religion are continuous and inseparable. My preliminary interactions with HESCO coordinators have indicated these similar overlaps among their dharmic values and scientific projects.

Although physicist Varadaraja V. Raman (2012) notes the easier availability of Indic ancient sciences such as astronomy, medicine, and mathematics as a positive by-product of Western Enlightenment, one of the fiercest critics of modern science, Vandana Shiva (1988), argues that "modern science takes into account only those properties of a resource system that generates profits through exploitation and extraction; properties that stabilize ecological processes but are commercially non-exploitative are ignored and eventually destroyed." According to Shiva, "As a system of knowledge about nature, reductionist science is weak and inadequate; as a system of knowledge for the market, it is powerful and profitable." Shruti Kapila (2010) summarizes the difference between the emergences of science in India vs. the West:

> The presence and persistence of a powerful and systematic rational tradition, both Hindu and Muslim, facilitated the deepening and refashioning of India's ecumenical tradition. In other words, this was a tradition that was incorporative in its approach to new ideas.[1] By contrast, practices of exclusivity defined the boundary-drawing exercise of much scientific and professional activity, which was then territorialized as separate scientific disciplines in the West. Such exclusivity had few, if any, existing parallels in the Indian context, where knowledge was accumulated and aggregated rather than hived off into competing sections.[2] By the mid-nineteenth century this ecumenical tradition was reformulated in the public, though competitive, new world of the print media.[3] This arena proved to be productive for debating the fundamental nineteenth-century question of the relationship between science and religion. However, the relations between religion and science in Europe and India were mirror images of each other. The emergence of science in Europe was an Event, in that it was a rupture in the preexisting arrangements between knowledge, religion, and authority broadly construed as the Enlightenment tradition. The Event of science was not constituted simply by its ritualized contestations over disciplinary exclusivity; rather, the specific eventuality of science in Europe was ultimately constituted by a confrontation between man and God. Whether this involved his "death" or his "exile," science had led, despite the dissenting tradition within the Enlightenment, to a categorical disenchantment with God.[4] By contrast, in India science was no Event.

In the light of this critique, the ecological regenerative philosophy of HESCO and of Baba Seechewal should thus be categorized into alternative science. Raman (2012) argues that some postcolonial thinkers "seem to be unwilling or unable to recognize that new ideas always transform cultures and civilizations, whether from internal or external impetuses, for the good in some ways and for the bad in other ways, that such transformations cannot be averted in the global-knowledge context, and that the alternative to change is stagnation in ages whose glories are more in the recalling than in the reliving". Thus, instead of rejecting modern science altogether, Raman argues for accepting that the scientific method and Indic traditions are not antagonistic. This almost matches the declaration by Hans-Georg Gadamer, cited in Dipesh Chakrabarty (2000): "Europe . . . since 1914 has become provincialized . . . only the natural sciences are able to call forth a quick international echo." Yet Amit Prasad (n.d.) argues that such universalization of science can also result in science becoming insular and provincial – a largely Euro-American and Australian intellectual engagement with a sprinkling of "others."

However, my own observations about HESCO match those of Seth (2009:378) in his comprehensive literature review of postcolonial science studies. Seth cites Akhil Gupta and others to conclude,

[C]onflicts or disagreements between "local" or "indigenous" forms of knowledge and "Western," "global," or "universal" science, particularly those where universal science seems at best ineffective and at worst profoundly destructive, are counter- productive.[5] Anthropologists (in particular) have evinced skepticism towards the very idea of an authentic, systemic, and autonomous "indigenous knowledge" as something to be opposed to a singular "scientific knowledge." "Local knowledge" cannot be simplistically equated with "indigenous knowledge".

Further, Seth cites Adams and Stacy Leigh Pigg, "Locality is socially and historically produced in and through a dynamic of interaction. The local is not a space where indigenous sensibilities reside in any simple sense,"[6] but exists as a hybrid phenomenon incorporating both "indigenous" forms of knowledge and "Western" sciences. As a Sanskrit verse in an ancient Indic agricultural text *Kashyapiyakrishisukti* mentions (Kashyapa et al. 2000),

As the time changes the king should take into account a change in the manner and mode of agricultural technique for sowing of seeds, and also consider the application of agricultural knowledge as different for cool and warm climates. . . . Cattle, rainfall, water reservoirs, and many other factors also cause this change. Accordingly, the king should manage the farming activities, depending primarily on the quality of the soil.
(II.168–171)

The methodologies of HESCO and of Baba Seechewal take into account the newer technologies, changing climatic conditions, and reviving the traditional ecosystems and cultures. In the next chapters, I demonstrate some of these hybrid and alternative models of scientific and sustainable models. As Dr. Joshi mentioned to me,

> In my childhood, I used to avoid brand-new clothes, shoes, or toys, and instead preferred to wear used clothes and shoes of my older brother and ride his used bike. In my HESCO work also, I never preferred to enforce a new technology. Instead, we strove to upgrade and revive the older traditions and technologies. Thousands of years of natural and cultural resources should be maintained using traditional technologies otherwise we will end up destroying them.

Critiquing the modern zeal for efficient technologies and arguing for a balance of physical labor and luxury, he said,

> Perfection should not be desired from anybody, even God is not perfect. Climatic climax leads to downfall. I am also not running after the results. God made everybody incomplete or imperfect which motivates him or her to put efforts to achieve perfection. A society that runs after perfection or efficiency is bound to downfall because this efficiency comes at the cost of environmental resources, and physical, mental, and spiritual wellbeing. This efficiency is not too different from luxury. Anxiety, heart problems, and sugar are all human created diseases because of this rat race. Balanced and multilateral thinking is required for sustainability. A society that does not do physical labor is destined to death. Physical labor is as essential as Oxygen.

Speaking directly on dharma and science, Joshi said,

> Dharma provides the rules to discipline the society such as by instilling the fear of God. Such rules also help purify and pacify one's body, mind, and soul. It makes one a better person. However, a true religious ritual (*puja*) should be a kind of Yoga in which nothing is asked from God. Rituals and worship should build your character helping you do better social service. There is no other "fruit" of puja. Dharma is not different from science. Dharma is not against the food and science shows ways to earn and cook the food. Science was intertwined with life and was present in every era but it was not categorized separately. All dharmic rituals and practices have scientific basis as well. For example, lighting a lamp in the Hindu dharma is done with revering the five great elements (*Punch Mahabhutas* of the earth, water, fire, wind, and space). Science gives the same importance to these five elements. Both dharma and science are complementary. Religious leaders

have defined dharma very narrowly. Science is not against the tradition. Any new scientific discovery used by society for a long time eventually becomes part of the tradition, culture, and dharma. Similarly, dharma and karma are the two sides of the same coin. In ancient times, we gave four different duties to people depending on their abilities and qualities but over time, they became caste-labels in their corrupted form. I never read any Hindu religious texts and even read Gandhi quite late in my life. I wanted to chart my own path instead of being biased as somebody's follower. Religious preachers of scriptures don't awaken the society properly. The first step to transform a society is to make it self-dependent. Service for others will naturally follow. We should leave them to their own karma (actions). Himalayan people now are enthusiastic about their resources and this is just the beginning.

Like most of the scientists described in this chapter, Dr. Joshi's views synthesize the Indic dharmic and cultural traditions with the latest science. In the following chapters, I present details of this synthesis.

Notes

1 Muzaffar Alam and Sanjay Subrahmanyam, *Indo-Persian Travels in the Age of Discoveries* (Cambridge: Cambridge Univ. Press, 2007).
2 Sheldon Pollock, ed., *Literary Cultures in History: Reconstructions from South Asia* (Berkeley/Los Angeles: Univ. California Press, 2003); see esp. the essays by Pollock, Muzaffar Alam, and Sudipta Kaviraj.
3 Seema Alavi, *Islam and Healing: Loss and Recovery of an Indo-Muslim Medical Tradition, 1600–1900* (Basingstoke: Palgrave Macmillan, 2007).
4 Akeel Bilgrami, "Occidentalism, the Very Idea: An Essay on Enlightenment and Enchantment," *Critical Inquiry*, 2006, 32:381–411.
5 Bruno Latour, *Science in Action: How to Follow Scientists and Engineers Through Society*, Cambridge, MA: Harvard University Press, 1987.
6 Vincanne Adams and Stacy Leigh Pigg (eds), *Sex in Development: Science, Sexuality, and Morality in Global Perspective*, Durham, NC: Duke University Press, 2005, p 11.

3 HESCO

An overview

Having heard about HESCO, I contacted their office near Dehradun, the capital city of Uttarakhand, which was carved out as a new state from the Himalayan districts of the north Indian state of Uttar Pradesh, to start my research in the summer of 2012 with the support of the Fulbright Environmental Leadership Fellowship. Later, I arrived at their office and soon visited various villages where HESCO has implemented several socio-economic projects. In several weeks that I spent with the HESCO staff and volunteers, I learned a lot about the perspectives of Dr. Anil Joshi, a retired Botany professor, regarding the society and the environment. What I present below is based on my several such interviews with HESCO coordinators and villagers working with HESCO. I have also extracted relevant information from the Hindi pamphlets and brochures of HESCO that is based on the writings of Joshi and his team.

Dr. Joshi, founder of HESCO, has repeatedly emphasized that the main goal of HESCO is to transform the rural society based on the foundations of dharmic principles and scientific technologies. For Joshi, dharma is not to be interpreted as religion, which can strengthen the social barriers instead of removing them. Instead, Joshi exhorted his students and coordinators to work for the empowerment of the Himalayan villages based on the idea of self-reliance and ecologically sustainable development. This is the impetus behind much of HESCO's work that aims for the rural sustainability in India.

As I have shown in my earlier work about other rural communities (Jain 2011), HESCO projects can also be interpreted from a Dharmic lens, not in the religious sense but from the ethical and environmental perspectives. A kaleidoscopic term such as dharma seems suitable to understand and interpret the kaleidoscopic work of HESCO. Like other rural communities such as Swadhyaya and Bishnois that I researched in my earlier work (2011), HESCO people also do not identify their work as a religious work but the way HESCO is trying to connect people with the environment reminded me about the similar work Gandhi and others have envisioned in which the culture and environment holistically sustain and support each other. Let me now present a more detailed survey of various HESCO projects.

Figure 3.1 Dr. Anil Joshi outside his cottage

Socio-ecological work of HESCO: a brief overview

Since 1980s, HESCO have striven for integrating the latest scientific and technological knowledge with local environmental and developmental issues in the Himalayan villages. For their pioneering work in sustainable development and its results in the rural areas, they have been recognized numerous times by national and international agencies. HESCO staff and volunteers not only live and work with the remote villages across Himalayas, they also encourage the local villagers to join in the constructive projects that aim to sustain the humans and nature. Based on this collaborative work with the locals, different ecological interventions are determined that are then applied in different areas to reduce the extreme poverty and resource crunch in some of the regions. Dr. Joshi and some of his colleagues established HESCO for the environmental conservation, under the firms and societies Registration Act in 1983. In 1991, they registered HESCO also under the Foreign Contribution Regulation Act (FCRA) in the year 1991 to be abled to receive donations from the international agencies.

However, the work of HESCO began as a local ecological movement when a young Botanist, Professor Anil Joshi, started taking students and social activists for the fieldwork in forests and mountains in Northern Uttar Pradesh region (now Uttarakhand state). This group later emerged as the Himalayan Environmental Studies and Conservation Organization (HESCO). Today, HESCO continues to discover and develop technologies that are suitable for rural areas. It also continues to seek collaborations with other international groups and organizations that are working in the rural areas in an effort to develop an international alliance of such groups across the world. It had developed several models of sustainable technologies that can be applied across the world.

The economy of the Himalayan villages in India is often termed "money order economy." It began when the majority of male members from the villages were recruited in the British and later Indian military. Today, this trend continues with men moving to Indian urban areas where they end up working on low wages. Sometimes, even women and children migrate and barely survive on handful earning and extremely poor living conditions. HESCO continues to attempt to reverse such trends and provide the helpful tools and technologies to enable farmers to remain grounded with their ancestral land and even be productive contributing members of their respective villages. Let us take a look at some of such programs.

Dharmic philosophy and scientific experiments of HESCO

HESCO has developed a new idea *Shridan* to promote cooperation and collaboration among different villages. As HESCO brings sustainable developmental ideas to one village, it encourages it to adopt a different village to share the knowhow widely. This promotes volunteerism, develops the local skills, and encourages self-sufficient solutions, with a priority on environmental protection and economic independence, using science and technology as a tool for sustainable development.

Watermills (Gharats)

In the Independent India, eminent Gandhian thinkers such as Vinoba Bhave and P. C. Ray continued to highlight the spinning wheel (*charkha*) and other indigenous technologies representing "an easy, healthy, natural process of increasing the wealth of the country and smooth way of universalizing the incidence of wealth" (Dharampal 2000, 53). But, most of the later scientists such as Meghnad Saha advocated for large scale scientific and industrial progress planned and funded by the State planning (Dharampal 2000, 53). However, many Himalayan villages were not even electrified until 1980s. To help fulfill the dream of socio-economic development, HESCO developed and deployed thousands of small turbine-like wheels to generate the power in the thousands of continuously flowing rivers in the Himalayan rural areas. This model of the revised watermill (*gharat*) proved to be an affordable and ecofriendly means to generate electricity in a large area of Northern India. HESCO manufactured these watermills using an affordable and simple technology that are now widely used in grinding flour, dehusking, and other local farming and household applications. About two thousand watermills were further upgraded during this period for various applications across Himalayan zones, especially in the border areas near China and Pakistan. In order to promote it further and to bring self-sustainability, a group of unemployed youths were trained to serve at the stations for watermills, and the motivated youth also ran a workshop on "farm fabric" to fabricate the turbines and other equipment for the mill. Such workshops provide service for installation repair and maintenance of the mills as well. Another important task undertaken is to develop new turbines for efficient output. These turbines were designed to cater to the power needs of other important local resource-based enterprises. A few new turbines have been designed to enhance output. These turbines are essentially locally fabricated with community participation.

Spring recharging

Another major area where HESCO has successfully intervened is in making fresh water available in remote mountain regions of India. It successfully started harvesting water in several older water bodies that were neglected and were not being properly used. By collaborating with BARC (Bhabha Atomic Research Center), HESCO utilized an environmental isotope and started discovering the water sources in underground springs in various Himalayan regions. As of March 2013, they succeeded in increasing the water flow in sixteen ponds and check dams in their mountain springs that has tremendously helped the nearby villages. The work has also started in nine more similar areas that will potentially benefit about 150 villages. About 120 springs are further treated for recharging. A water-recharging laboratory has also been developed with the support from BARC (Bhabha Atomic Research Center). The laboratory is a facility center for villagers to help them identify exact catchment areas of a spring for treatment, with a capacity to analyze samples for 200 springs annually.

Worth from Weed

Lantana is a weedy plant invading forest and community land in the Himalayas. HESCO's intervention to control the weeds took an interesting turn when it subjected such species to local and commercial utilization. HES-CO's slogan is *Khar Patwar Ko Khara Karo*, i.e., turn the weed into something valuable. Instead of having to burn lantana, HESCO developed it as a viable material for designing household furniture. Several household items are now made from it, and other NGOs have also started working on a lantana cottage industry. HESCO organized training sessions to demonstrate using the leaves from local trees such as neem (*Azadirachta indica*), bhimal (*Grewia optiva*), and lantana for mosquito repellents and making charcoal from the waste biomass using pyrolyzers. HESCO devised the techniques to harvest it for the production of furniture, construction materials, beehives, and storage tanks for grain, fodder, and water. Its leaves may be used with caution for incense sticks and scented candles, since it contains phenylethanoid glycoside. Many household items continue to be developed using the weed.

Women's Initiative for Self-Employment (WISE)

WISE is an attempt to answer the various needs of women, especially in the rural parts of the Himalayas. WISE believes that appropriate technology for women's upliftment is important, especially in this region, where they are engaged in all household and agricultural activities. The focus thus became skill development to enhance the potential of women and to equip them for the future. This gave birth to the development of low-cost, appropriate, location-specific, easily adopted technologies especially for the women's employment. Currently, WISE has 1,120 members, 340 of whom have developed their own enterprises with an average turnover ranging from less than one to two million rupees annually. The present turnover of WISE (with the group of individual entrepreneurs) is forty million rupees annually. Another task that WISE undertakes is to encourage the women in the villagers to develop their skills for economic contributions. Using the "post-harvesting technologies" for value addition activities, e.g., many women are involved in making the bakery products that are sold across the country.

Farmers Bank (Kisan Bank)

The resentment of Indian farmers increased in the recent past, leading sometimes to suicide or severe depression. Three major factors decide farming fate, namely farming services, the role of banks, and price determination for farm return. HESCO has developed its own Farmers Banks (*Kisan Banks*) with ten branches across the state of Uttarakhand.

Technology Initiative for Peace (TIP)

In addition to collaborating with scientists and policy makers, HESCO has also worked with the Indian military to develop localized technological solutions in Himalayan regions, especially the strategic border areas near Bangladesh, Bhutan, Burma, China, Nepal, and Pakistan. Some of such projects include connecting Indian villages in these regions with the electric grid using the watermills and also providing livelihood from the locally grown fruits and other crops.

Youth employment

HESCO has also initiated a dialog with the ministry of home affairs and the ministry of youth welfare to involve youth of rural India, especially from naxalism-affected areas, in economic development activities based on local resources. The youths from thirty districts of eleven states of India have been trained to develop local employment opportunities, such as branding, marketing, and increasing productivity.

Religious offering as employment

According to HESCO's estimates, various ritual offerings (*Prasads*) into millions of places of worship in India is a large multimillion cottage industry dominated by outside business entities that are non-native to these pilgrimage sites. To help local communities take advantage of this business opportunity, HESCO is training them to prepare offerings from local natural resources. This is helping local communities find new avenues of employment and also better recognition for local resources (see Figure 3.2).

Traditional agriculture

HESCO is reviving traditional crops in the mountains, which are better suited to the local ecology and human physiology and are less intensive to cultivate. HESCO has reviewed these crops in more than 200 villages, and different varieties of inputs and better cultivation practices have been reintroduced, such as buckwheat and finger millet. As I observed during my field work, the villagers as well as the shrine boards continue to receive HESCO sweets made from locally produced grains.

Work with the Dalit communities

The Dalit communities continue to be discriminated against and inhabit the border areas of the villages of the Indian Himalayas. These communities are

Figure 3.2 Women packing religious gifts for distributing in a temple

deprived of many development advantages. In 2007, HESCO worked with the local governments to select thirty-five Dalit-majority villages across the country to develop their own labors and skills, which were upgraded with new scientific and technological inputs.

Dharma in HESCO

HESCO and other similar Indian NGOs continue to revive many other long-neglected technologies and small-scale industries. It should not be surprising if at least some of them (with minor modifications here and there) prove to be as productive and cost-efficient as the new technologies, which Indian policymakers have borrowed from modern world industry (Dharampal 2000). Upon my questioning of Dr. Joshi about dharma and HESCO, he quipped,

> Food and water is part of the dharma. Our work is rational based on scientific technologies. But I did not want to force dharmic practices on others. I slowly became more religious and god-fearing. But I inspire my colleagues to do their *karma*. Once they are relieved from my leadership, God will inspire them. In Gandhi's times, people were more religious. But today, society has changed a lot. The nature will soon be considered God. We clean our river on every Saturday and that is our

puja. In Gandhi's time, nature was not so polluted and people did not need to worship the nature. A common man does not analyze about God rationally and I regularly remind them that your natural resources are your direct and manifested gods and you should not disrespect them. Belief in God should reflect in service the poor, not just rituals in the temples. Sanskar and Sanskriti are synonymous with tradition. A scientific experiment when practiced long term becomes a part of Sanskriti. *Gharat* today is Sanskriti and so is a pressure cooker and phone. HESCO does not ignore the traditions. In marriages, cookers are gifted. When the first time, it whistled, the old lady ran away from the kitchen. But today, it is an essential part of life. Similarly, the *gharat* when first produced the bulb, people were shocked. Similarly, the springs are recharged to revive the old traditions: "My village, my water." The catchment has to be repaired and revived.

Joshi's words, as recorded from my interviews with him, speak for themselves pretty clearly and navigate effortlessly from the realms of dharma and science repeatedly. For him (and his colleagues), there is virtually no distinction between science and dharma. He continued,

HESCO does not want to be obsessed about dharma. Joshi never read a lot of books. Gandhi used to say the same what HESCO is doing. We want to be original and unbiased for any ideology or tradition. *Punarjanma* (rebirth) is used as a concept and as an ideology to plant trees in the memory of diseased family members in Himalayan villages. However in our other projects, for selling the villages' products, the priority has to be quicker sale and that is why an easy name or brand is created and used without solely depending on Hindu terms. Moreover, we want to be secular towards all religions instead of preferring any one kind of names. All religions talk the same: serving, respecting others, rules etc. They all want to create an ideal society. Dharma can be means but if you completely depend on dharmic language, you might divide the society into smaller segments (Hindu/Muslim etc.). HESCO's objective is the same as of all religions – to create an ideal society. All the religions want everybody to be independent so that they can serve the society. Otherwise, one will continue to be selfish. . . . My family has only reluctantly supported my work. I take lesson from Gandhi and Kasturba's lives. I don't want to force my wife or anybody else to imitate or follow my work. It should be voluntarily chosen. So, I discuss about those people who have participated in my work more than my family. HESCO work is like a *yagya* and there should be no evil thought or sinful act here.

Again, Joshi is careful not to depend on the Hindu side of the dharma too much to keep a safe distance from the Hindu right. Instead, he prefers to model his ideology and methodology based on Gandhian ideals of *Sarva-Dharma-Sambhav*, i.e., equal regard for all the religious traditions. At the same time, Joshi is careful not to fall into the Gandhian mistake of forcing one's family members to follow one's chosen ideals. I noticed that none of the positions in HESCO has been given to any of Joshi's family members, and Joshi continues to take great pride in his colleagues in a spirit of a larger HESCO family instead of any trace of nepotism in or around HESCO.

4 HESCO

A brief history and ethnography

Foundations of HESCO in the Himalayan state of Uttarakhand

In Uttarakhand, the main occupation in the villages is agriculture. Since most men have been migrating to the cities or work for the military, women perform most of the duties related to their farms back home. The Himalayan states are often referred to as the states on the "Money Order Economy," i.e., the local economy depends on the money sent by its migrated men in the cities. According to a HESCO survey, in addition to employment opportunities, other reasons for migration are poor education and medical facilities in the Himalayan villages. It is often mentioned that there is a lot of land in these villages but not enough people to work on the land.

The land is divided into small pieces dispersed in remote mountains. About 15% of this state is irrigated; the rest of the land is dependent on rain. In addition to wheat and rice, people grow millet, lentils, fruits, and vegetables. This diversity in their crops helps them survive against droughts, pests, and insects. Most of the farming is done using organic methods. Most of the biomass is utilized for fertilizer, fuel, and fodder. The entire state is filled with forests, streams, and mountains, which provide several necessary ingredients for humans and animals. Although in recent years cities such as Dehradun have seen a massive loss of trees, the *Himalayan Gazetteer* described the state as prosperous and perennially green. Both the Garhwal and Kumaon regions of the state were fully self-dependent and were more developed than other states of India (Atkinson 2002).

In the Himalayan region of Uttarakhand, about 24 kinds of fruits, flowers, buds, and leaves are used for human consumption. The animals consume about 15 kinds of fodder. Although forests and soil are losing their natural qualities, as of now, life in these areas still largely depends on natural resources. There are about 142 different kinds of cuisines that are made from the resources from the forests and agriculture that grow about 32 kinds of vegetables, 9 kinds of spices, 12 kinds of grains, 10 kinds of lentils, and 8 kinds of oil-seeds (Singh 2001). With this background of the Uttarakhand's agriculture and ecology, in this chapter I trace the chronological path of HESCO's inception with brief biographical notes of its founder Dr. Joshi and his colleagues. I hope these interspersed introductions of HESCO folks will enrich the historical description of HESCO with my ethnographic work with them at dozens of Himalayan villages. All the information presented here is based on my interviews with dozens of HESCO coordinators and other villagers in Uttarakhand and Himachal Pradesh.

HESCO's founder and leader, Dr. Anil Joshi, was born into the lower middle class in a small town called Kotdwar in the North Indian state Uttar Pradesh; the town is now part of the new state Uttarakhand, which was carved out of the northern part of Uttar Pradesh in 2000. Recounting his early life, Joshi shared with me that he experienced resource scarcity early on. His mother's industriousness left a life-long impression on him. Joshi has four brothers and two sisters. He and his siblings were deprived of several basic childhood pleasures, which were readily available to other children in their neighborhood, e.g., a bicycle, new clothes and shoes, etc. His father worked as a day laborer and even as a tailor. Academically, Joshi was not a favorite among his teachers, had to struggle hard to pass high school, and with great difficulty managed to get his admission into the college in Kotdwar (now named the Dr. Pitamber Datt Badthawal Himalayan Government Postgraduation College; I visited in June 2012 and met its current principal, who went to great lengths in describing Dr. Joshi's work there in 1970s and 1980s). Joshi's life took its first positive turn when he started taking his education seriously and became a good student in his college and soon completed his bachelor's and master's degrees and earned a doctorate, all in botany. In 1976, immediately after earning his PhD at age 25, he succeeded in becoming a lecturer at the same college in Kotdwar and started teaching graduate students. His academic zeal continued as he aggressively continued his research agenda in the local forests and published several research papers. He also earned his reputation as an extremely strict teacher. These were also the years when the Chipko Movement was emerging as a very successful movement in the Himalayan region. In 1981, Sunderlal Bahuguna was given the Padma Shri Award by the Indian government, and in 1982 Chandi Prasad Bhatt was given the Magsaysay Award. The issue of the environment also captured Joshi's imagination and he joined the movement led by Bahuguna.

Continuing his life story, Joshi shared that he married in 1984 with great reluctance. He had started having an intuition that he would accomplish great social work. Apparently, he had already decided that his life was meant for the larger good of society, therefore he wanted to sacrifice his personal interests. Only after a lot of persuasion from his mother, he agreed to the marriage. After marrying, his mother once took him to a renowned astrologer who predicted that Joshi's fame would be compared with that of Mahatma Gandhi. The astrologer also predicted that Joshi would launch his own school after renouncing his job at the local college. Joshi feels that almost all these predictions came true, and he never felt unsafe about his life as his confidence was only verified by these forecasts. According to Joshi, the fear of God disciplines society. Only humans have the power to think, which connects them with God, and only humans have the ability to serve others, even at the sacrifice of their own interests. We can see the glimpses of Joshi's dharmic and scientific perspectives here.

In 1979, within a couple of years after starting his teaching career at Kotdwar College, Joshi began the work of bio-fencing (fencing the farms to protect from wildlife using native plants), basket designing (using local weeds and other plants), and mushroom cultivation in the local villages. Joshi was given a separate space from the botany department where he created a new laboratory and a nursery. In 1980, he helped launch the first fruit-processing unit in Kotdwar. In 1982, he conducted his first march to raise awareness about natural resources, and in the same year, the Indian government's department of environment gave him a project, with thirteen of his doctoral students, for studying grassland ecosystems in the Western Himalayas. This was one of the earliest and largest projects for the study of biomass in Garhwal. The project was completed in 1986 and the report was submitted to the department of environment in New Delhi. This was his first major step from conducting laboratory-oriented research to society-oriented research. In hindsight, the project can also be seen as one of the earliest pedagogical attempts in India to embrace societal service in regular classroom learning (now called "Service Learning" in the United States). Again, we see glimpses of Joshi's efforts to transcend the boundaries of dharma and science in his research and teaching.

Recollecting his early turning points, Joshi told me that around 1981–1982, once Joshi visited Ghad, a nearby village, to raise awareness about forestation. During his speech, a local woman suddenly interrupted him and challenged him to solve their basic problems, such as education and health, instead of lecturing about saving the forests. Joshi recalls this as a pivotal moment in his life that changed the direction of his research and activism forever. One of the questions that Joshi has raised ever since is why should the people living in the mountains act like guards of natural resources when these resources are taken away for the consumption by others? Why should these villages be treated just as a market for the finished industrial products or providers of the raw materials for these products? As he put it,

> Instead of mass-production, we want to encourage production by the masses. We want development with equity. Earlier, a village depended

on its natural and human resources, such as barber, water miller, farmer, blacksmith, and cobbler. The village economy was based on the barter system and resources were conserved for thousands of years. Now, all the products are centrally manufactured in the cities leading to the degeneration of village economy and ecology. This has led to the collapse of sustainability. The skill of the masons, blacksmith, and carpenters must be upgraded so that they can be productive again. Decentralized economy is the panacea for the environment and for the economy. Centralized and vertical economy is the root cause of many problems. In this system, energy is first wasted in transporting the raw material and then again via PDS (public distribution system), finished products are brought back to the villages leading to more energy consumption along the way. Also, there is an unaccountable energy with the migration of rural folks to cities leading to chaos and crowded slums there. This model creates economic disparity and eco-logical losses. This model requires more input and produces less output and so is not sustainable. HESCO has discouraged aggressive agricul-ture resulting in more equitable input and output. Unfortunately, in the modern India, service-based economy is dominating the productivity-based economy and HESCO is against this. Youth should be inspired to move towards productivity-based professions rather than service-based. Journalist: keep a safe distance from Joshi lest you might join his work, he has great motivating power! It is not important how good you are. Important is, how many people you can influence. If you are really good, show it in how many people you can benefit.

HESCO launched with "planting of thorns"

To continue his social and ecological service in a more organized way, in 1983, HESCO was formally registered.[1] HESCO's first office was started at Joshi's paternal home, which HESCO's volunteers used both for office space and sometimes for staying. Started at the family house, everybody joined Joshi like an extended family. Joshi continues to command a great respect from his colleagues, many of whom are also his ex-students. Like his house, Joshi's scooter is also shared as a common property, and in my observations I noticed that HESCO folks continue to live like family mem-bers sharing their clothes, shoes, bags, and suitcases.

Comparing HESCO with other NGOs, Joshi mentioned that HESCO is not an NGO but a family in which all personal matters of each member are jointly discussed. Unlike an official mission that an NGO might declare for itself, HESCO has kept its goals and missions open to all the local social needs and issues. Once an Indian government officer suggested to Joshi to own a car. Joshi politely refused it, stating that all the great social movements in history, such as by Gandhi, have worked without such luxuries. HESCO also does not want to leave its carbon footprint on the land for which it strives to work. Although all HESCO members now use mobile phones, it was only in 2006 that Joshi accepted it as a necessary technology and not just a luxurious one. Recalling his experiences with other NGOs, Joshi shared his astonishment upon visiting the Vivekananda Center in Chennai, where he saw luxurious cars and buildings. According to him,

> HESCO's focus and the real resources are the villages, therefore HESCO does not need to depend on any external resources. The luxuries should not be part of any social service. The social workers should not charge for what they share, such as their knowledge or other kinds of services. This is why some activists tend to prefer speaking at elite venues and charge hefty fees. But trees like these are destroyed by just one storm because their roots are very weak. They are not grounded on the firm soil of villages. This is the reason why many NGOs are now tarnished; they are run like other businesses. Like Gandhi, one should renounce everything step by step and then one will be able to work for the people. Gandhi's was a voluntary movement, now it has become NGO business. Why should I tie up with any NGO, it is not a business! Their ideology may be well articulated but their way of working is corrupted. Once I visited the Cheetal Grand Hotel in UP and the waiter refused to allow me in looking at my clothes. I immediately protested and was shocked to see a poor insulting a fellow poor. My turban often becomes a reason for my insult.

In 1985, one of HESCO's current collaborators, Prakash Joshi (who also heads the Himalayan Organization for Protection of Ecology) graduated from Govind Ballabh Pant University of Agriculture & Technology at Pantnagar and then went to Lucknow. One of his relatives forwarded his information to HESCO and he soon started working at Kotdwar on an eco-development project at Ghad village in Pauri district in 1986. Based on its success, HESCO was awarded a core project from the Department of Science and Technology (hereafter DST) for the next five years. At the successful completion of this project, he was sent to Kumaon. Based on his knowledge about his native village in Kumaon, he concluded that lack of education is a major problem. He decided to launch an elementary school at his village. As he shared with me, the village happily gifted him a piece of the land and his wife sold her jewelry for additional support. On August 15, 1993, he inaugurated this school and I visited it during my fieldwork in June–July 2012. Continuing his story, he told me that he has merely provided a lamp for his village and only the people can supply the necessary fuel for its successful operation in the future. Prakash Joshi, trained in modern college education, ensured that English is also taught at this school from the beginning. Recounting the challenges in enrolling enough students, he told me that he had to provide candies and biscuits for the kids, get their hair and nails cut, and get their clothes cleaned. Such personal attention helped in creating a bond with them, and the school has been running successfully ever since. In addition to language study, resource education is emphasized. Later, he started a high school in the same vicinity, some of whose alumni are now well placed in different positions as engineers or doctors. In 1995–96, DST awarded a fisheries project to Prakash Joshi that was highly successful. He designed a large fishpond costing only 25,000 rupees and motivated many villagers to donate their labor for it. He also worked on a project on the local Himalayan plant *rambans*, as well as on a water and sanitation project and diversified agricultural project supported by the World Bank. In Loharkhet, a wasteland was donated to Prakash Joshi, on which he has developed bamboo and herbal plants. From all his projects and community work, he continued to win several awards and a great reputation in the surrounding villages in the Ranikhet area of Kumaon. Prakash Joshi, however, deferred this honor to his mentor Dr. Anil Joshi, whose influence was apparent in his sociocentric work and simple lifestyle. Recalling one instance, Prakash Joshi told me that once he had lost 900 rupees, but Dr. Joshi immediately reimbursed him without any question. What we learn from this brief biographical note about Prakash Joshi, a HESCO partner for last couple of decades, is that the dharmic-social-scientific philosophy of Dr. Joshi has trickled down to HESCO's other partners also. Science or other knowledge or technology in itself is of little value for them unless it is used for social upliftment and personal transformation, such as avoiding addictions or other luxuries.

Resuming the HESCO story, the organization continued its work on the study and conservation of Himalayan biomass, landslide control, and distribution of medical supplies in the local villages until 1988. The HESCO team afforested about twenty-one landslide sites and has continued building several check dams using locally available stones. During the final portion of the rainy season, seeds are broadcasted for helping the afforestation. According to Dr. Joshi, natural means and local resources should be used to stop natural disasters, instead of importing solutions or resources from the outside. With regard to stopping the landslide, HESCO found several local plants that could survive even with the sliding land in the heavy Himalayan rains, as had been the case since ancient times. Joshi's faith in local traditions and resources such as these continue to be strengthened throughout his work, as we will see later in this chapter. Joshi asked his team of students to collect seeds of many of such plants and plant them at strategic places to deter the landslides. Several kinds of grasses were also taken from different areas and replanted elsewhere for testing their utility in stopping the landslides. Eventually, *rambans* was selected as the most suitable plant. Because of this successful project, HESCO gained recognition for the first time, and society and the government started paying attention to their work from then on. The Border Road Organization (BRO) also joined HESCO for clearing the roads and stopping the landslides, and about five thousand BRO workers planted *rambans* according to HESCO's directions. About one thousand students of Joshi had already joined in this project. According to Joshi, he succeeded in motivating the people because he took up issues pertaining to their everyday lives. At the time of my fieldwork in June 2012, the Indian prime minister's scientific advisor was working with several Himalayan NGOs under HESCO's leadership to stop the landslide by aggressive afforestation. Other major research institutions of India, such as the Central Building Research Institute (CBRI) and Defense Research and Development Organization (DRDO), are also working with HESCO on this project.

One of the other early innovations by HESCO was the gasifier that used the local biomass for its fuel instead of imported diesel that would have been expensive both financially and environmentally[2] (Agarwal 1995). In 1989, the Indian government rewarded HESCO's work in a major way by including it among twelve NGOs in the entire country. This was a major success for HESCO, as this list is selected after a very carefully scrutiny of the NGO's work and even in 2012, there are only seventeen organizations listed as core NGOs by the Indian government's Department of Science and Technology (DST).[3] One of the early core projects was about the study and conservation of the kandali fiber plant, another local Himalayan plant. Following my timeline, it is time for another biographical pause to introduce one of the earliest students and co-founders of HESCO, Dr. Rakesh Kumar.

Rakesh Kumar, who completed his PhD in 1989 under the supervision of Dr. Joshi, has continued to co-direct HESCO ever since. Upon my asking Kumar of his work with HESCO, he started talking about his "guru" Dr. Joshi and his social work. In 1989, Joshi conducted a foot-march from Kotdwar to Yamkeshwar to spread environmental awareness among the local people. Dr. Joshi sponsored the education of several students, once rescued a Muslim girl from domestic violence and helped her open a shop to repair umbrellas, and helped several people from alcoholism. In another instance, a stranger named Rohtas came to Dr. Joshi's office pleading for saving his wife. Joshi took them to the hospital and convinced the doctor to treat the woman for free. Later, Rohtas, who was from a "backward" caste, joined HESCO and was trained to ride a bicycle. Rohtas soon quit smoking and made a pledge to gift a car to Joshi if ever he wins a lottery! Similarly,

Joshi helped a woman named Sarita marry her divorced husband, helped another woman named Bishveshvari Devi's daughter get married and start a fruit-processing unit. There are a number of such instances of Joshi helping several people socially and economically throughout the HESCO story, and Joshi's charisma continues to inspire his colleagues and other people around him. It is time now to introduce another key HESCO coordinator, Dr. Kiran Negi.

Dr. Kiran Negi is one of the senior scientists associated with HESCO since its inception. Her bachelor's degree in science was from Tehri Garhwal. For her master's degree, she joined Kotdwar College in 1988 and then earned her PhD in agroforestry in 1995 under Dr. Joshi's supervision. Joshi used to teach ecology and would take his students on Sundays to the mountains to work on stopping the landslides, not different from recent service-learning-based courses now becoming popular in American classrooms. They also used to publish booklets on these issues. She was initially interested in research and work towards her PhD. Joshi motivated her to chart a research career that is based on applied research with and for society, instead of on purely theoretical "armchair" research. Joshi had already published several research papers but had realized the futility of this exercise, as it had little impact on society. In 1990, Dr. Negi registered for her research on mountain grasses for a project in partnership with Jawaharlal Nehru University in New Delhi. She worked on Himalayan fiber plants such as *rambans*. HESCO discovered that this viviparous plant would be ideal for stabilizing the soil and also for fencing against aggression by elephants, because viviparous plants produce seeds that germinate before they detach from the parent resulting in a quicker growth. Negi also trained farmers about mushroom cultivation. Continuing my journey of sketching brief biographies of key HESCO personnel, let me mention about another early female colleague, Vishweshwari Devi Tiwari.

Tiwari's mother was the head of local women's group *Mahila Mangal Dal* (MMD, a women's grassroots group across the Himalayas, which had played an important role in the Chipko Movement and in earlier Gandhian movements). However, after Tiwari's marriage, at her in-laws home, she was not finding any avenue for social work. She and her husband were unemployed, so they went to the nearby village of Pakhi, where there was once a conflict between the two upper-caste groups of Brahmins and Kshatriyas. Tiwari suddenly found an opportunity to engage with her new neighbors. She visited each family to reconcile and unite them and soon emerged as the head of the village. Later, she was elected to be the head of the local MMD. She met Dr. Joshi when he was visiting Gopeshwar for a meeting in 1989. He was impressed with her energy and invited her to work together with him. Tiwari went to Kotdwar for a two-day workshop led by Dr. Kiran

Negi, then Kiran Rawat). Later, Rawat held a meeting at Pakhi with Tiwari and other women of the ten nearby villages. According to Rawat, the area and its people were familiar to her, as she had grown up in that area. She expressed her interest in working in these villages. The villages also had a legacy of the Chipko Movement, as the famous Chipko leader Gauradevi was from one of the nearby villages. People asked HESCO for their basic needs, and the organization was interested in resource management. Choosing the middle way, Rawat conducted a survey of all the villages and then designed a program. Four women came to Kotdwar for training and stayed with Rawat and other HESCO folks. HESCO later installed the Bhekal oil extractor and distributed medicines.

In 1992, Rawat and her colleagues shifted to Pakhi itself. Fortunately, her parents always allowed her to go and work with the HESCO team. It used to take twelve hours to reach this village from Kotdwar back then. HESCO motivated the women to form an organization so that villagers could sustain this work on their own without depending on HESCO permanently. HESCO also started creating centers for fruit processing, and Dr. Joshi used to visit every month.

In 1992–93, gasifiers were installed in Tiwari's villages Pakhi and Paini as an alternative to lack of electricity. Later, based on her suggestion, Dr. Joshi invited the state chief secretary of Uttarakhand, Dr. R. S. Tolia, to Tiwari's village to see the locally designed products based on a Himalayan "dwarf bamboo" *ringal*. Impressed by her leadership, Tolia recommended Tiwari for the village head position and she was eventually appointed. HESCO arranged for the training of village women in juice making, mushroom growing, sewing, and weaving. According to Tiwari, other NGOs offered them cash but did not give them work and yet were jealous of HESCO's involvement. Later, the newly trained women, led by Tiwari, imparted training to women of other villages. Local natural products such as apples, potatoes, and mushrooms were chosen for processing and production. However, according to Tiwari, many such training initiatives did not lead to actual business units, although models were still being created for profitable farming. Eventually, Tiwari shifted to a bigger village, Nagarasu, for a couple of years in search of a better school for her kids. In 1998, she was supported by HESCO to register for her own NGO, called the Himalayan Environment Conservation Women Society (*Himalayi Paryavaran Arakshi Mahila Society*). In 1999, after a major earthquake in Chamoli, the Department of Bio-Technology (DBT) awarded an agriculture and fruit-processing project to Tiwari and HESCO provided her the funding. Commenting on gender problems, Tiwari mentioned that the local men protested a lot against women earning and gaining socio-economic independence. Earlier, the village head used to give us permission whether we can go out or not, but they started maligning our names. Now, they have come to respect Tiwari and other women of their villages. Traditionally, in Himalayan villages, a woman's day starts with waking up quite early, cleaning up the cattle, and then taking the tools to the forest for fodder collection in the season or collecting firewood. Due to their work with HESCO, men were concerned that this traditional role of women would diminish. But as Tiwari put it,

Today village women, their daughters, and daughters-in-law have stood up for their rights. People insulted me a lot but with the blessings of Joshi Sir, we have shown to everybody that illiterate poor women can also achieve their goals. All that is needed is persistence and intellect. Even educated people hesitate in expressing or taking any stand but we can do all kinds of tasks. We are not limited to cleaning our cattle but we can do everything. Even food inspectors threatened us as soon as we started earning in our NGO. We also go to extinguish forest fires, even kids come along with us. I am no longer the head of the village but still perform all my social service, even more than before.

Another woman of the same village, Lakshmi Devi, joined HESCO in 1993. She had completed her eighth grade education but "had never seen a train" (a common Indian expression to denote one's rustic background). She went to HESCO for training in sewing. HESCO arranged for her trip, on a local train, to Najibabad. Even in early 1990s, this part of India had very few transport facilities. Lakshmi Devi, in turn, trained other women in her village, following the standard HESCO model of spreading the knowledge around. Lakshmi Devi also led a small project to develop a nursery on a wasteland and planted radish and cauliflower with cow dung as the fertilizer. Once, their village was under threat from blasts due to road construction by the state government's Public Works Department (PWD). Dr. Joshi intervened and convinced PWD not to conduct blasts that were causing cracks in local homes and throwing debris onto local farms. Dr. Joshi also met the local district magistrate (counterpart of an American city mayor) and received a written assurance to stop the blasts. Dr. Rakesh Kumar noticed Lakshmi Devi's food-processing skills and invited her to HESCO's new center in Nagrasu. She later married another HESCO colleague, Shersingh Rawat, and then shifted back to Paini. As of 2012, carrying her own equipment, she continues to visit village homes to train others.

Another woman of this village, Kamladevi Bisht, told me that eleven women from Paini earned 25,000 rupees from selling *Prasad* sweets to Badrinath (a major Hindu pilgrimage center of Uttarakhand). This group maintains a small community-based bank to save and loan to each other. HESCO sent Bisht and thirty-four other women to training about farming and fruit processing. Organized by the M.S. Swaminathan Foundation at the University of Agricultural Sciences, Dharwad, and conducted by Professor Nirmala, this training equipped them in turning local materials such as millet and other grains and fruits into desserts such as *Halva* and *Kheer* (pudding), as well as making cakes, biscuits, and other kinds of snacks. The training also included designing vermicompost bins. Traditionally, millet used to be thrashed by using legs, so all the trainees were also provided with millet-thrashers, which are more convenient and efficient. These trained women later visited the Govind Ballabh Pant University of Agriculture and Technology at Pantnagar and shared their newly acquired skills with about thirty families there, as well as with women in Himachal Pradesh. Such early training activities had to be organized along the caste lines that tended to be blurred over the period, and current training classes need not be categorized along upper and lower castes. Let us now turn to one of the earliest and long-time HESCO coordinators, Prem Kandwal.

Kandwal used to work for the forest department in 1989 when Dr. Joshi was on his foot-march about the issues of anti-alcoholism and forest preservation from Kotdwar to Yamkeshwar. Kandwal also joined the march and became familiar with Dr. Joshi's philosophy and colleagues. After Kandwal completed his three-year contract, he approached HESCO and started working at Ghad village on landslide control with Prakash Joshi, whom we met earlier in this chapter. Kandwal also worked on lantana (considered an invasive species or noxious weed) and other agricultural projects in Kotdwar after Prakash Joshi left for Kumaon to work on his school project. The lantana project was co-directed by Dr. Joshi and a couple of professors from other major research institutes of India. In this project, lantana was discovered as a mosquito repellant. It was innovatively recommended for making furniture for the first time by HESCO,[4] which became a source of income generation for hundreds of people. In 2010, Mahseer Conservancy, an NGO on the periphery of the Jim Corbett National Park, joined hands with HESCO to market lantana furniture with the brand Corbett under the project *WELFARE* (Women empowerment through Lantana Furniture and Artifacts and Restoration of Environment).[5] According to Dr. Joshi, lantana developed across the Himalayas due to forest mismanagement. HESCO connected it with the local economy that eventually stopped its growth. HESCO also recommended to the government agencies that remote sensing should be used to explore lantana's locations. HESCO slogan is *Khar Patwar Ko Khara Karo*, i.e., turn the weed into valuable. Instead of having to burn lantana, HESCO developed it as a viable material for designing household furniture. Several household items are now made from it, and other NGOs have also started working in the lantana cottage industry. In 1991–92, a training session was organized at the Dr. Yashwant Singh

Parmar University of Horticulture and Forestry in Solan, Himachal Pradesh. The training session demonstrated using the leaves from local trees such as neem (*Azadirachta indica*), bhimal (*Grewia optiva*), and lantana for mosquito repellents and making charcoal from the waste biomass using pyrolyzers.

Resuming HESCO's chronological story, in 1995–96, Dr. Joshi fought against the forest department when it started demanding penalties from people for cutting lantana. Later, UREDA (Uttarakhand Renewable Energy Development Agency) took over the project to manage this plant. Bhimal is another fiber plant known as the "*pahad ka peepal*," i.e., peepal, *Ficus religiosa* or sacred fig, of the mountains (peepal is one of the most commonly worshipped trees in Northern India, see Haberman 2013). Joshi urged the government to make a special department for bhimal on the lines of southern states' initiatives for sandalwood and coconut. Bhimal is an all-purpose tree; its leaves are used as the green fodder for animals and the stick is used as fiber and for generating light for walking across the Himalayan villages in the dark hours of night. Bhimal is widely found across the Himalayas; approximately six to eight trees of bhimal are found around every farm, especially in the lower altitudes. HESCO evolved bhimal to be used for household items such as mats, bags, hats, and even for shampoo (but later shampoo was discontinued to avoid involving chemical usage in it). Several training sessions were organized at Kotdwar, Rishikesh, and elsewhere. With HESCO's activism, today there are several bhimal and *rambans* (agave) fiber utilization centers across Uttarakhand. Similarly, Bhekal was upgraded for extracting its oil, and several other local aromatic plants were upgraded to make incense sticks for religious rituals. Continuing the list of biomass innovations by HESCO, it was discovered that mushrooms grow well on straw, which is not readily available, so one of the HESCO workers, Prem Kandwal, worked on identifying the waste biomass for mushroom cultivation and emerged as an expert on mushrooms. Over the years, he has helped hundreds of people utilize every little space in and around their homes to grow mushrooms. HESCO developed several other wild and non-conventional fruits by grafting. For instance, the stem of another Himalayan tree, mehal, was grafted with pear, resulting in quicker yield, faster growth, and stronger plant. Similarly, different varieties of apple were also grafted.

As part of its *Jal, Jangal, Jameen* (Water, Forest, Land) initiative, HESCO also developed the seed banks (*Kisan Bank*) for seed preservation by and for farmers. Seeds are made available to day laborers through major research institutions such as the Indian Council of Agricultural Research at Almora and the Govind Ballabh Pant University of Agriculture and Technology at Pantnagar. Local Agricultural Science Centers (*Krishi Vigyan Kendra*) also provide help and support to HESCO's Kisan Banks.

After their crops, farmers return these seeds to the Kisan Bank to be shared with other farmers in HESCO's sharing spirit, called the *Shri Dan*. There are three "call centers" to help farmers that are managed by the professors at Pantnagar and Kotdwar. Since 2009, community-based banks are also operating in ten districts (counterparts of American counties) with a single window system.[6] Such micro-financing facilities are especially useful for smaller farmers, as also noted by the state minister of Uttarakhand at its inauguration.

In 1991, in the wake of the major earthquake in Uttarakhand, Dr. Joshi invited Laurie Baker (d. 2007[7]), the renowned British-born Indian architect and humanitarian who had pioneered the design of earthquake- and tsunami-proof housing.[8] Baker joined hands with HESCO to build the earthquake-resistant houses using local resources. Baker later called Dr. Joshi his "mountain son."[9] In 1999, HESCO built more such earthquake-resistant houses as well as twenty-two community sheds.

Let us now meet another HESCO collaborator, Raghuveer Kandwal, who founded Rural Reformation and Labor Service Institute, aka GRASS (*Gramin Sudhar evam Shramik Seva Sansthan*), in Mayali village in Rudraprayag District in 1991 with Dr. Joshi as its patron. In 1991, Dr. Joshi came to his village after the earthquake and later called him to Kotdwar. Joshi advised him to focus on the fruits commonly available in his village. Kandwal told me that malta grows during winter but they did not care to preserve it. With HESCO's guidance, they started preserving it and kept adding more people in the process. HESCO organized several meetings and workshops and also introduced Kandwal to more people. With HESCO's support, Kandwal also installed watermills and started producing electricity locally. HESCO also helped him design fishponds and install grain-processing machines. Today Kandwal coordinates about 150 local self-help groups' activities such as fruit preservation, seed banks, and fishponds. They collect malta from local farmers and help them in marketing their products such as jams, squashes, ketchups, and sauces. The peels are used for making powder and the seeds are used in nursery. Recalling his experiences with the government's horticulture department, Kandwal told me,

> They (government trainers) used to come and train us based on their pre-determined calculated agenda. But we never followed it up after the training was over. HESCO's training process was more adoptable and easier. Today, 20 out of 100 villages are connected with us. Once we invited Joshi Sir for dinner and served all the Punjabi dishes but he instead wanted the local Garhwali dishes. He instantly connects with the local people because he speaks and lives like us. This is why HESCO's approach is more successful than the government projects. Joshi

Sir also sent us to the Central Food Technological Research Institute in Mysore and to Govind Ballabh Pant University of Agriculture and Technology in Pantnagar for training. Joshi Sir makes a phone call and even an illiterate villager like me can enter the top research institutions. But he would never use his name for personal motives. I recently bought a car because it is really useful in conducting my work here with 5,000 people but am hiding the car from Joshi Sir. A government agency CAPART had gifted a Gypsy car to Joshi Sir in 1995 but Sir had returned it in 1999. His scolding keeps me away from greed. I regard him as my father. I had once asked for a computer but he rejected my request like a father.

Another crop, soybean, grows abundantly in this part of the Himalayas, and farmers used to barter it with the traders in exchange with salt, jaggery, or rice. With HESCO's intervention, they stopped bartering it and turned it into a cash crop by selling it to the cities directly. HESCO provided them some new techniques and also sent them to a training workshop supported by the World Bank. Today, farmers are turning soybeans into milk, snacks, tofu, and other processed food items and selling them both locally and to the cities. The processed fruit products are sold to herbal companies such as Hamdard and government agencies such as Garhwal Development Corporation Limited (*Garhwal Vikas Nigam Limited*). During my fieldwork, I visited a Kisan bank, a fruit-processing unit, and a beauty parlor, all managed by local women. Their combined turnover has grown from 5,000 rupees to 1.5 million rupees over a period of one decade. These improved socio-economic conditions have resulted in better education for local children, who have now started going to local schools and even colleges in other towns.

Continuing his memoir, Kandwal told me that in 1996 another earthquake hit and the need for a quake-resistant community hall was felt, so five community halls were built with attached toilets and water tanks. When a government agency, the Council for Advancement of People's Action and Rural Technology (CAPART), proposed temporary shelter-camps for earthquake victims, Dr. Joshi instead argued for permanent shelters that should be earthquake resistant. Later, the Department of Bio-Technology also joined HESCO's project and provided 2.5 million rupees, 250,000 each to ten sheds. Villages contributed the unskilled labor and land for these shelters. These sheds were built based on the HESCO philosophy of using only local resources, which are more cost effective and durable.

In normal times, schools are run in these sheds. I also learned that a Christian missionary approached Dr. Joshi, but the missionary absconded with half a million rupees, which were contributed by eleven different NGOs.

In this village, I also met two teenage girls leading their self-governing councils *(Bal Panchayat)*. They have filed about 50 petitions using the Right to Information Act, a new law passed in 2005 allowing any Indian citizen to request information from a government office. Having received media training, they were publishing a local newspaper and a magazine. They had worked on the issue of child rights, birth registration, and child marriages. This is what I wrote in my notes as I interviewed them:

> We made a report on discrimination against a girl child which we presented at the children conference at Uttarkashi and called it "As We See It." Kids from all thirteen districts came. Then we presented it at a conference in Delhi organized by Plan International. Then we represented India at a United Nations conference in Uttarakhand. We also went for a seven-day trip in Geneva with kids from fifteen other nations. We met Human Rights Commissioner and also interacted with many children from other countries.

I also met Govind Singh Negi, the headmaster of the local high school, who has been connected with HESCO for the past fifteen years. Recollecting HESCO's involvement with local villages, he told me that HESCO motivated them to use leather products, such as jackets and gloves, made from dead bodies using low-cost technology. HESCO has worked hard to revive the old traditional professions such as masonry, blacksmithing, and leather working. Similarly, traditional Ayurvedic healthcare was revived, and Ayurvedic doctors were affiliated with the Herbal Research Institution *(Jadibuti Shodh Sansthan)* directed by Dr. Rakesh Sundariyal. Artisans were upgraded with new tools and training from the Indian Institute of Technology. According to Negi,

> Our ancient traditions need to be upgraded. Laboratory scientists create their theories just by their limited armchair theories. The real scientists live in villages; they just need to be promoted, encouraged, and upgraded, e.g., plough-makers, toolmakers, mason, and watermill workers. Mason constructor should remain independent and self-employed instead of becoming an employee to an outsider builder.

In 1992, HESCO coordinated collaborative projects with eight NGOs and a few scientists across the Himalayan region. This initiative helped these eight NGOs become teams with practical hands-on solutions rather than mere activists pressurizing government and other agencies for solutions. The State Institute of Rural Development (SIRD) in Lucknow organized training for NGOs for the first time that helped train hundreds of villagers.

Hopefully, the preceding few paragraphs gave at least a glimpse of the diverse range of work that has been going on with the inspiration and guidance of Dr. Joshi and his graduate students and also supported by the government offices at the center and the state. However, his academic colleagues did not appreciate this social work, and eventually Dr. Joshi's supervisors asked him to be transferred out of the Kotdwar College where he was still working. Instead of accepting this order, Dr. Joshi decided to resign from his teaching position to focus on HESCO work full time. To his credit, the Indian government's Department of Science and Technology rescued his job by putting on a deputation for a couple of years. As soon as he completed his 20 years of teaching service, he resigned from his job in 1993. According to Joshi, he wanted to remain a teacher and a student instead of working in the administrative committees. His family opposed his decision to resign, but he never felt insecure. His belief in God was only further strengthened by the way he kept getting the support throughout his journey from various directions, although a few local political leaders were initially against his efforts to save the forests.[10] Also in 1993, the prestigious Ashoka Fellowship was awarded to him, and his work was labeled as biomass utilization.[11] Accepting both challenges and rewards, Dr. Joshi kept his faith in dharma and conviction in science to continue working for the Himalayan communities.

Moving to Gwar Chauki

Continuing its journey, in 1994 HESCO completed a project to create a comprehensive inventory of all the natural resources of Himalayas. HESCO was now looking for a village where it could start its work without much financial investment, and fortunately a family friend named Meharban Singh Negi "Guruji" gifted his paternal land near a village called Gochar. On March 12, 1994, HESCO moved to Gwar Chauki (a village in the district of Rudraprayag) and started constructing a new campus there, which was later named as Vigyanprastha.

Initially, HESCO designed small cottages using bamboo, cement, and gunnysacks as well as an access road. Eventually, the Indian government's Ministry of Rural Development's CAPART provided funds to HESCO to develop a rural technology resource center. There were about seventeen families in the village that already knew Dr. Joshi because his father was from the same village. HESCO tried to relate with their new neighbors by attending their weddings, participating in their street plays, teaching their children, and even tolerating the initial protest by an ex-head in the village. Eventually, children started going to school and today are well placed in the Indian military and elsewhere.

Initially, people thought that Joshi, who used to be referred as Dr. Joshi, must be a medical doctor, so they started asking for medicines in exchange of their farm produce. Recognizing their health needs, HESCO established a health bank and issued health cards to all the families. Doctors from Indo-Tibetan Border Police (ITBP) were asked to visit monthly. As of June 2012, an Ayurvedic doctor (Dr. Bhawani Dutt Maithani) was employed at the campus, and I visited his office during my fieldwork. At this new center, HESCO experimented with several different kinds of technologies, which could potentially have both benevolent and malevolent uses. For instance, bio-pyrolyzers were developed to make charcoal from waste biomass, such as decayed pine leaves. HESCO presented this as a solution for wildfire, and the briquettes and coal thus produced were suggested for firewood and heat. However, some government forest officers expressed their concern that this technology may lead to villagers using forest resources aggressively for making charcoal. HESCO's response was that any technology has two potential usages: constructive or destructive. It is for the users to decide which way they want to go. Similarly, water-harvesting sites were constructed to provide water for wild animals[12] away from the village to minimize their invasions into the villages. Unfortunately, this also had the potential to be misused by the poachers who could easily target animals near these water tanks.[13] Villagers built these water tanks based on HESCO's ideas and they maintained these tanks themselves. They proved to be cheaper and eco-friendlier than plastic tanks available in the markets.

One of the prime needs in the village was sanitation. Most of the Himalayan villages did not have toilets, and people used to defecate on the roadsides. HESCO decided to start working on this with the support from the Department of Science and Technology. Within a few years, all the villages to which HESCO reached out were installed with toilet facilities. Initially, people were reluctant to change their defecation practices, but slowly toilets were accepted as a preferred option.

I learned about another instance demonstrating Joshi's philanthropic spirit. When HESCO came to Gwar Chauki, there was only one buffalo in the entire village. Realizing its importance for a family's economic and nutritional needs, Joshi wanted to give away 50,000 rupees that he had received from a grant. When he offered this money, some women rejected his help saying that there wasn't enough water necessary for the upkeep of the buffalo (which require more water than other animals for bathing). The water was so scarce that villagers used to take baths only occasionally. HESCO remedied this situation and developed several water tanks, and today every family in this village has a buffalo; milk is consumed locally and sold in the markets. In the beginning, about 50,000 rupees were distributed among four women. Later the money was shared with fourteen more families based on the concept of *Shri Dan*, i.e., money is distributed interest-free from one successful family to the other. According to one of the local farmers, Shishupal Singh, whom I met at Nagarasu,

> We got a lot of useful information and training in farming and rainwater harvesting from HESCO over last few years. We also received the repair funds for watertanks, seeds for vegetables and plants such as mango, guava, grapes, pear, ginger, turmeric, mulberry, jackfruit, and peach. We also received the Sugandha turmeric developed by the YS Parmar University useful in cooking and for making dyes. This has resulted in our financial gains. We try to share such information and resources with others.

HESCO also distributed mango saplings to every family on the condition that any plant lost will be penalized (so that proper care was ensured for every plant). Fruit processing was also developed with local human resources, local technology, and local natural resources. Today, farmers regularly grow vegetables that are both consumed and sold in the market.

In the mountain villages, castes are divided according to the forests and altitudes of the mountains. For instance, a girl living around the pine forest is not married into a village near another kind of forest because she is not used to carrying the fodder on her head. (Pine forest's fodder is carried on the back, not on the head, because pine trees grow in dry lands while trees such as oak ensure richer water sources leading to heavier fodder to be carried on the head.) Similarly, the Thakurs (warrior castes) live at the highest echelons, then the Brahmins (priest castes), and low-caste communities live at the lowest altitudes. Joshi challenged this system by deliberately choosing a low-caste person to cook for him. It was met with heavy resistance initially by the upper castes but they were all convinced eventually. Once, all the teachers were invited to see the HESCO technologies with a complementary lunch by HESCO. This was termed *Gyan Panchayat* by Joshi to discuss all the village issues. One of the teachers wanted to segregate himself because of his low caste. Joshi immediately decided to tackle this problem. In his kitchen, he replaced the upper-caste maid with a low-caste woman. When the village started protesting this, Joshi threatened to leave the village in his challenge to casteism and untouchability. With HESCO's continuous efforts, today the entire village accepts each other without any untouchability or casteism.

HESCO has also conducted a pioneering project in eleven different locations in India, supported by the Department of Science and Technology, in which the principal investigator was required to be a person of a low caste. Any upper-caste people working on this project were required to live with low-caste families. All eleven villages set an example of caste equality. At the culmination of the project, India's agriculture minister and 400 other people also dined at the low-caste village. Dr. Joshi asked all the dignitaries to contribute 500 rupees for their meal, and this benefited the village.

In Gwar Chauki, according to Joshi, he wanted to create a center in which all the nineteen families of the village should also have their ownership in addition to HESCO volunteers and employees. Dr. Rakesh Kumar, the current HESCO director, told me about a woman named Mangla Devi, who had returned to the village from her husband's home in the city. She did not have enough food in her home. When Dr. Joshi came to know about her plight, he immediately went on a fast as a punishment for his colleagues – how could a family be hungry in HESCO's presence in the same village! The entire team also joined in this fast. Eventually, Dr. Rakesh Kumar lent 500 rupees to Mangla Devi for buying groceries. She also asked for potato seeds and immediately started digging her farm in the hot summer month of June without waiting for even a plough. Dr. Joshi appreciated her enthusiasm and did not want to stop her. Eventually, she grew a good crop of potatoes and returned the 500 rupees. Later she attended the HESCO training for making pickles. From her savings, she constructed a new house and called her husband back to the village from the city where he had been working as a day laborer. After the family was settled well in their native village, HESCO invited them to share their transformation at a conference organized by the Department of Science and Technology. As of 2012, the couple was successfully working on their farm and their child is pursuing his undergraduate degree in business administration. They also bought a buffalo from another loan given by HESCO that they later repaid from the additional earnings.

Continuing his memoir, Dr. Rakesh Kumar told me about a man named Yashveer Negi whose father was working for Border Road Organization. Negi once mentioned that he is waiting for his father's death so that he can get his father's job (most of the youth in India find it extremely difficult to obtain a government job). Realizing his desire, his father approached Joshi, who arranged for the training for Negi. After the training, Negi started working as a plumber in the village. Later, when he was selected for a coveted government job, he turned it down and chose to stay and work as a plumber in the village instead.

At Gwar Chauki, HESCO also introduced the vermicompost and developed the idea of Kisan Bank further. As stated before, such banks were started to encourage the habit of saving money. With the faith in Dr. Joshi and his team, one of the HESCO employees, Shersingh Rawat, had accumulated about 800,000 rupees from the nearby villages that would be eventually deposited into a government bank. During my fieldwork, I met one of the farmers, Vikram Singh, who grows vegetables (such as okra, chiles, eggplant, onion, and green peppers) with the support from HESCO. His earning had increased manifold, enabling him to send his children to college. He had also built a water tank for rainwater harvesting and to feed the cattle. HESCO calls such added resources a model of integrating technology with farming.

Over the years, HESCO has found that growing vegetables is easier and more affordable than growing fruits. There is also less decay in vegetables than fruits. Hence, HESCO has designed several methods to add value to the fruits. These fruit-processing methods provide twofold help. They help preserve the fruits for longer and thus increase the profit margins of the farmers by reducing the role of middlemen. There are about 140 such units across the Himalayas as of 2012, some of which we already noted in the stories about Tiwari, Kandwal, and other HESCO coordinators and collaborators.

Another interesting initiative started in Gwar Chauki is called *Chulhon ka Sansar*, i.e., the world of stoves. HESCO has collected twenty-one different kinds of traditional cooking stoves and has displayed and given brief information about each model. Each stove is refurbished to minimize input and maximize output. In 2007, HESCO had supplied about 180 stoves to Vishnu Prayag. Farm Fabrics, an association of watermill owners trained by HESCO, fabricated these stoves. This is yet another example of HESCO organizing artisans and craftsmen.

Mushroom and vegetable cultivation was encouraged and a mushroom spawn preparation unit was established in 1995 at Gwar Chauki. The unit is used for seed preparation of mushrooms, which was used by Prem Kandwal in his work as we saw previously.

I also met Karan Singh (member of a "scheduled" low caste) who had founded the Women Upliftment Rural Development Council (*Mahila Utthan Gramin Vikas Samiti*) in Brahmakhal in 1990s. In 1995, he met Dr. Joshi at a hotel in Dehradun, the capital city of Uttarakhand, for a project funded by the Department of Science and Technology. He told Dr. Joshi about his NGO and soon Joshi assigned a project to him. Singh's NGO organizes training for about 900 women and a few men in different self-help groups. They also work on recharging water sources. Singh's NGO was awarded under the Golden Jubilee Self-Employment Project with the inspiration of HESCO. According to Singh,

> Any poor or downtrodden person can meet Joshi at any time day or night and is deeply impressed by his love, care, and simplicity. He changed my life; I could have gotten a government job but Joshi Sir inspired me to work for the society instead of running after a secured career. I am helping about 28 villages with 13,500 people. Once my

daughter got the admission into a college and I could immediately get 25,000 rupees for her admission from his help. Similarly, I had my appendix operation and I received help again from him. I have returned all the money back to him now. We need a couple of more people like him to project our state Uttarakhand as a model state to the world.

It is time now to introduce another senior HESCO coordinator, Shersingh Rawat, who had been with HESCO since 1995. Rawat joined me in my trip to the famous Hindu pilgrimage center Kedarnath and shared many interesting memoirs:

> Many times Joshi Sir helped me financially such as by paying for my child's treatment. Shivam (Joshi Sir's son) respects us like the elders of his family. We all dine and live together whenever I visit HESCO center in Dehradun. Whatever we work, Joshi Sir gets the report. He always encourages us happily. There is never any financial pressure in HESCO and all our expenses are taken care of. Joshi Sir always asks us to work with the villagers. He talks very congenially with villagers, dines and lives with them. During harsh winter times, he gifted me his own sweaters many times. In 1996, Balvant Bhai of village Gwar had a stomach injury and Sir helped him with 6,000 rupees. We are always ready to work on any task for HESCO because of our relationship with Joshi Sir. There is no differentiation or distance kept; it's like a family. I used to design all HESCO pamphlets, brochures, and fliers but now we outsource them. I had job offers in Delhi and Haldwani but I wanted to live in Garhwal, my native place and fortunately, I found my job with HESCO. Then Joshi Sir called me to Dehradun for four years and I started going for fieldwork. My diploma is in electrical engineering so that knowledge was also useful for the villages. Once HESCO organized a micro-technique workshop in which a toolkit is given to each unemployed youth. About 30 women from Rishikesh came for this training and I gave my first lecture in that workshop. Joshi Sir liked my work and I began working on such training projects more and more. I conducted training in sanitation, agriculture, fishery, beekeeping, vermicompost (which takes about 80 days for the making of fertilizer), NADEP compost (composting method developed by Naryan Devrao Pandri Pandey), and bio waste (such as grass, cow dung, and sand).

I learned from Dr. Rakesh Kumar that Rawat was introduced to HESCO by his relative teaching at a local polytechnic college. Rawat's real skill was later observed to be working with people, so he was diverted towards that work, although people are motivated to work in many areas. Rawat was HESCO's publishing expert as well as fieldwork expert. Later he married another HESCO worker, Lakshmi Devi, whom we met earlier. Rawat was also the key personnel in 1996 when HESCO launched its own journal called *TIME* (*Technology Intervention for Mountain Ecosystem*).

Resuming the HESCO story, in 1996, with training from National Institute of Design, HESCO trained villagers in making small baskets and lampshades from *ringal* (Himalayan dwarf bamboo). People of Rudia caste (a schedule caste) and a similar group in Uttarkashi with fifteen *ringal* workers took advantage of these training sessions. In the same year, Koti was developed as a model village (Kumar and Dobhal 2006). This village had twenty-five low-caste and twenty-five upper-caste families. HESCO researchers found that the process of feeding the fodder to animals was flawed. With HESCO's efforts, about 30% of the wastage in fodder was stopped with the help of the fodder machines given to forty-five families. Fodder was planted and fodder chopping was taught to the women. With this training, fodder's consumption became more efficient. Initially, *Mahila Mangal Dal* (women's self-help group) assumed that the animals were not eating enough, but later realized that the wastage was stopped, not the actual consumption. HESCO's approach has been to respect the local tradition and if possible to upgrade them for more efficiency. At Koti, HESCO built toilets and bathrooms for forty-five families; each toilet cost 2,500 rupees, about a quarter of the usual cost. Pathways, a nursery, a fishery, sewage treatment, soak pits, a community hall, water tanks for rainwater harvesting,[14] and a playground were built and two traditional watermills were upgraded. Self-help group (SHG) meetings have been held here regularly. With more efficient dairy management, a loan of about 50,000 was returned to the dairy department in 2011. Following the Koti village's successful model, several neighboring villages have emulated construction of many of these facilities.

Moving to Dehradun, the capital of Uttarakhand

In 1996, Joshi's son Shivam fell ill, and recognizing his chronic condition, Joshi decided to move HESCO near a big city. Also due to lack of communication and other facilities in Gwar Chauki, HESCO shifted to Dehradun in 1995–96 in a rented room. In 1999 near Dehradun, HESCO designed the Women's Technology Park (WTP) with DST's grant of half a million rupees (there are only three or four such WTPs functioning across India). WTP played a key role to connect the women of Himalayan Indian states of Jammu and Kashmir, Himachal Pradesh, and Uttarakhand. As Kiran (Rawat) Negi, the main coordinator of WTP, mentioned,

> There are at least 35 species of fiber-yielding plants in the mountains. We taught the women to use the fiber to make things that are generally made of coir. It also offers a two-week training to make pickles,

chutneys, jams and squashes using local fruits and vegetables. HESCO has got village elders to identify medicinal plants and herbs, and now they are grown in nurseries in many homes. It has also identified 65 plants whose leaves, barks, flowers, fruit and root can yield dyes. Women are taught to use the dye and make marketable items like pouches, file covers and spectacle cases. All the women trained in the Women Technology Park later become members of Women's Initiative for Self-Employment (WISE) network to promote and market their produce.

Dr. Joshi mentioned this about the Himalayan women,

As a child, I was pained by the drudgery of the people, particularly women. They had to walk long distances to bring water. And they had to leave home at an unearthly hour to go into the forests to ease themselves. Today many villages around have water harvesting tanks and Indian-style toilets with septic tanks for each house. HESCO has also devised a multipurpose kit named Kalyani, comprising a tiny screwdriver, cutter, and copper tweezer instead of the crude gadgets women used to carry around to trim nails, clean their ears, and remove the painful thorns as they trudged through the forests.

Jagatram Ramol lives in Nahan, a town of Himachal Pradesh on the border of Uttarakhand, and has been running his NGO called YAMDHA since 1985. As of June 2012, Ramol was coordinating the spring recharging projects in collaboration with HESCO and the Bhabha Atomic Research Center (BARC, India's most prestigious nuclear research center in Mumbai). In 1996, Dr. Joshi conducted a meeting in Sirmaur that Ramol had hosted. In 2000, Dr. Joshi led a foot-march from Paunta Sahib, a town between Nahan and Dehradun, to Shimla, the capital of Himachal Pradesh. In this march, all the water resources were surveyed, covering four districts and their surrounding areas. A memorandum was given to the Chief Minister Dhumal who came to the final event of this rally. Ramol was very pleased to share with me that he had received a reward of 10,000 rupees from Dr. Joshi for successfully conducting the rally and the final event. Ramol was also HESCO's key collaborator in a few other DST projects that helped about twenty other NGOs. He was also the coordinator for HESCO's farmers movement (*Kisan-Kisani Andolan*). For this, two bicycle rallies were conducted, from Vaishno Devi (a famous Hindu pilgrimage center in the Himalayas) to Dehradun and from Kanyakumari (the southernmost tip of India) to Dehradun. According to Ramol, this campaign succeeded in motivating thousands of farmers to pledge against selling their land to any outside business. Instead, farmers were advised to demand a share in any business being planned on their land. Recalling Dr. Joshi's informal approach, Ramol told me,

> In 2000, Joshi Sir called me on phone and drove from Dehradun to Nahan in two hours. He came and asked for the local Himalayan bananas. He also gave me the posters and banners for the upcoming farmers campaign. Joshi Sir once also called me for a DST meeting at Dr. Y. S. Parmar University of Horticulture and Forestry at Solan, HP.

The only anthropologist working at HESCO is Sudha Sharma. She joined HESCO in 1997 as a research scholar and completed her PhD afterwards. In June 2012, I met her at a local elementary school in her office, located near Dehradun. Her niche area of work within HESCO is alternative education for the children. She also established a group of children selling fruits and vegetables on roadside.[15] She shared with me that once Dr. Joshi joined her in visiting a village Pali in Tehri district to gift bats and balls to children. She also publishes a newspaper for children called *Bachchon Ka Akhbar* (*Children's Newspaper*).

The paper publishes information about local natural resources to share with the children spread across hundreds of villages in Northern India. According to Sharma, modern education dissociates children from their village, culture, and resources, and her effort is to supplement the regular curricula with activities and knowledge about local natural resources. HESCO participated with the Indian national planning commission and gave its recommendations on education. Some of the ideas shared were, "The education must connect the children with their local resources. Skills should be developed to associate students with their resources. Education must make one independent, not dependent on others for employment."

Resuming the HESCO story, in 1997, Joshi led another march from the frontier village Mana (located on the India–China border near Badrinath) to Maleth. In 1997, from November 2 to November 26, he led *Jalandolan*, a water movement from Gangotri to Delhi. This was a 300 kilometers (about 187 miles) long march dedicated to water issues. The procession had a human chain for about nine kilometers (about six miles). Dr. Joshi met three state-level ministers during this campaign and apprised them of water problems in the villages. As he told me in one of the interviews, India's longest river Ganges originates in Uttarakhand and with its various tributaries, these rivers cover about 22 kilometers (about 14 miles) of the state. However, out of 16,000 villages of Uttarakhand, 12,000 suffer from water scarcity. Ironically, local villagers have very little access to the rivers flowing past their villages, and only three possible usages of water that they perceive are suicide, funeral, or hydropower projects being developed by the government. According to Dr. Joshi, his campaign was to demand better local access to water for drinking and irrigation. He also highlighted the plight of all the tributaries of Ganges and Yamuna, instead of only focusing on the Ganges in the plains, as is sometimes done by the government's environmental efforts. With the deforestation in the Himalayan regions, underground water has also reduced, although there are still sixteen *dharas* (rivulets) in the Gauchar area with better water quality than the World Health Organization (WHO) recommendation. Around 1999, the government wanted to involve HESCO for its Svajal Water Project for laying the pipelines for drinking water. However, HESCO denied it for three reasons. First, the scheme appeared unsustainable in terms of its resources. There was also a potential for conflict among the villages for the access to the water. Finally, an outside profit-making corporation was going to control the entire project. Although HESCO helped Raisira village in Pauri Garhwal, where water was supplied from a source located 10 miles away after Joshi approached the government officials, eventually HESCO focused on recharging the natural springs, which I will describe later.

In 1999, HESCO received its first major national award, called the Jawaharlal Nehru Prize. Fifty years after India's independence, in 1998–99 the Indian Science Congress Association instituted the annual award that used to be given to NGOs for the popularization of science. The award carried a cash amount of 100,000 rupees and a plaque.

In Dehradun, I met Kala Bisht who had received training for ten days from HESCO 1999. She used to grow strawberries, which used to rot very fast. She used her training to preserve her fruits and set up a center for herself. She (and twenty other women) later went to Mysore, Mumbai, New Delhi, and Switzerland for more training (facilitated by HESCO). Bisht later imparted this training to others.[16]

In 1999, Dr. Joshi visited Uttarkashi as part of his watermill movement and met Ramesh Anthwal, who was a reporter with the local Hindi newspaper *Amar Ujala*. Later, Anthwal founded an NGO called Rural Agriculture Development and Environment Development Institute (*Gramin Krishi Vikas Avam Paryavaran Samvardhan Sanstha*) with the support of HESCO. Anthwal, with HESCO, has built rainwater harvesting tanks and fisheries and has worked with artisans and craftsmen of various skills. From their estimate, about 80% of water tanks are working successfully across twenty-four Uttarkashi villages. They also participated in the United Nations' World Food Programme and distributed seeds to local farmers. When I interviewed Anthwal, as usual, he started talking about Dr. Joshi and called him "a good fearless administrator." Recalling his projects, he told me,

> Blacksmiths were organized to make new tools such as ploughs to be sold for cash and the earlier barter system, in which they used to

provide services in exchange of groceries, was stopped. Models of plough were taken from different agricultural universities. Exhibitions were organized in Uttarkashi and the tools were displayed at local fares for sale. Similarly, in Chauniyad village, people had stopped growing their traditional crops such as lentils and millet so we developed integrated projects for cattle, goat keeping, poultry, kitchen garden, fodder training, vermicompost, and nursery. The hens were eating the worms so hens had to be caged.

Another HESCO collaborator, Jagdamba Prasad Maithani, with an undergraduate degree in science, graduate degree in sociology, and a diploma in mass communication and journalism, founded Alaknanda Ghaati Shilpi Federation (AGAAS)[17] in 2000. Its governing body and its advisory board list about twenty people. They work with forty-five self-help groups with their eleven full-time employees. Maithani was a key coordinator during Dr. Joshi's Gangotri to Roorkee foot-march. After the Chamoli earthquake of 1999, AGAAS worked on a HESCO project, with support from DBT, for the quake victims. They raised two "fodder proof" nurseries (i.e., with plants that are inedible for cattle or monkeys; monkeys do not attack the okra, chile, or turmeric plants). One of these nurseries still generates revenue in Pipalkoti and is registered with the horticulture department, which buys all its plants, but the other could not be monitored properly due to its remote location. AGAAS also built two earthquake-resistant sheds for the community's common shelter in Chamoli, which is currently used as a high school. CAPART funded it and HESCO coordinated it. As of June 2012 when I had met Maithani, AGAAS was developing rural employment projects such as fruit processing, designing *ringal* products, growing organic pulses, and organizing ecological tours for students and trekkers. AGAAS had also received support from UNDP (United Nations Development Programme) for designing a bio-tourism park with low-cost cottages. AGAAS was also promoting the five lesser known trekking routes with the help of the Indian School of Business (ISB), Hyderabad; in 2011, seventy students participated in this project. Mumbai's Ratan Tata Trust helped them in another project to extract a special shrub, nettle, which is used to make yarn. Nettle's newfound quality has caused its price to rise from 15 rupees per kilogram to 60 rupees. Baskets are made from 36 hectares of *ringal* planted under The Mahatma Gandhi National Rural Employment Guarantee Act (MGNREGA), although about 35% of the area was destroyed from the wildfire. AGAAS was also HESCO's key partner in organizing and celebrating the Himalaya Day on September 9 in 2011 and 2012.[18] They conducted 88 training programs between 2008 and 2012 for 2,500 *ringal* artisans; 162 of them work full time while others work on seasonal projects.[19] Uttarakhand government has now declared *ringal* cultivation a small-scale industry, and traditional rights of the forest-dependent community were identified. On October 1, 2010, a permanent order was released that a community which permanently depends on the forest for *ringal* extraction is allowed to extract it for its livelihood, and the forest department will try to regenerate *ringal* instead of penalizing the community for its extraction. According to Maithani,

> This was a big milestone for us. Schedule Tribal (ST) communities weave and make yarn and Schedule Caste (SC) communities extract nettle and *ringal* because they do not hold their own land. STs hold two pieces of land in higher altitudes and in lower altitudes as well. I also worked for anti-liquor campaigns but wasted five years due to false allegations. I also helped organize "social army" to keep the villages clean and encourage afforestation in response to a request from some women. STs used to make local liquor but 16 villages stopped its sale by our activism. However, I was kept in the prison for 22 days due to false allegations of assault, which could not be proved because I was out of town at the time of the accident and I was released. Even inside the prison, I started a movement against injustices. I was disillusioned with HESCO when Joshi decided to go back to Dehradun, the capital. I wanted to live in the villages and continued to work in Pipalkoti which I did with my own NGO AAGAAS.

Resuming the HESCO story, in 2000, Joshi led a march from Paunta Sahib to Shimla. In 2001, Kiran Rawat was awarded a national award of 100,000 rupees for women's development through the application of science and technology, but this money was also deposited into HESCO's account, as is the HESCO tradition of using all prize money for its projects. In 2001, Shri Ram Washshran Devi Bhatia Memorial Charitable Trust awarded the Social Science Award to Joshi.

In 2001, Women's Initiative for Self-Employment was launched. WISE works with the local market instead of looking for remote markets. Processed food items, such as pickles and jams, are consumed locally. They are developed using natural local raw materials, and local farmers themselves sell them as WISE products. Processed food items and "nutri-bakery" items, such as millet biscuits and cakes, have been highly successful because of their fast consumption in local markets, such as the pilgrimage centers at Vaishno Devi, Badrinath, and Kedarnath.

I also met Dwarika Semwal, whose farm and shop were destroyed in the 1994 flood and 1997–98 landslides. Semwal was in the twelfth grade then. Being born in a Brahmin family, he was only familiar with traditional lifestyles but was intrigued to explore an alternative career choice after meeting Dr. Joshi in 2002 in Uttarkashi. Semwal was impressed with his thoughts and simplicity and wanted to work with him. In 2003, after the death of his parents, all the household responsibility fell on him. He went to Dehradun to tell Dr. Joshi about his interest in the nursery work. Dr. Joshi sent his colleague Dr. Rakesh Kumar with 10,000 rupees. Semwal told me about this turning point,

Those 10,000 rupees changed my life forever. I started a nursery and sold crops for 35,000 in the first year itself, it was a great beginning of my new life. We started a tree plantation in association with local schools to stop the landslides. This was the beginning of a new journey. Then we formed a new NGO "Jaddi," Himalaya Environmental and Herbal Agro Institute (*Himalaya Paryavaran Jadi Buti Agro Sansthaan*). We organized several meetings and plantation campaigns. In my nursery, I had fodder trees and fruits such as malta and mulberry. I planted fewer trees to make sure that they survive properly. I wanted to be a scholar and asked Joshi Sir how can I become one? But he advised me to stay away from such people who have maligned the social sector by their baseless theoretical knowledge. He suggested that I should stay in the village and awaken the power to help others without depending on external resources. I followed his advice and also traveled with him throughout India and learned from people. In 2004–05, from Lakhamandal to Lamb village in Uttarakhand, I participated in a foot-march to study the people's living conditions affected by the climate change. In 2006, when Joshi Sir announced about the bicycle rally from Jammu to Dehradun, I learned bike riding within one night at the IIT Roorkee campus but received bodily injuries. Joshi Sir and two dozen of us took a train from Dehradun to Jammu. We went to the temple of Vaishno Devi and after lunch, were ready for our rally. I checked my bike and was pleased to be second in the rally behind Joshi Sir. However, being a novice bike rider, I hit many people including Joshi Sir. Then Sir advised everybody to take care of me. Near Chintapurni in Himachal Pradesh, it was an uphill ride and my tire burst. I had many such challenging and learning experiences in the entire rally. After this rally, we designed small and large rainwater harvesting tanks. We organized blacksmiths and *ringal* workers. And then came another bicycle rally from Kanyakumari to Dehradun in 2008. My son was just born before the rally but I left him with my wife and her parents and joined the great opportunity to see the entire country firsthand. After the successful completion of the rally, I was honored by the district officer of Uttarkashi and now I regularly get invitations to participate and lead the campaigns related to the environment and development. [see Figure 4.1]

Figure 4.1 Dr. Anil Joshi leading a bicycle rally

Semwal told me that he was invited as an advisor for the firefighting campaigns. He was also an advisor for a project in which people working under Mahatma Gandhi National Rural Employment Guarantee Act will work with local *Van Panchayats* (local forest councils). He worked for the Yamunotri Prasad project as well, in which local people and local natural resources were utilized for the temple offerings. On May 5, 2011, a meeting was held to discuss the plight of the Yamuna River, and it was decided that twelve villages would clean Yamuna weekly with local people taking the lead (*Jiski ladai uski aguvai*). In Baagi village, Semwal helped launch a science village with a millet bakery and wheat processing. A pyrolyzer was also installed using teak leaves to make charcoal, which were not useful for fertilizer. Such projects helped to generate local employment and stop the wildfire.

Reviving traditional watermills (*gharats*)[20]

In 2002, Dr. Joshi received another national recognition when the popular English magazine *The Week* selected him to be Man of the Year.[21] Also in 2002, he led a march from Chokta to Dehradun on water issues. Soon *gharat*, watermill, was declared as an industry by Vasundhara Raje, then union minister for Small Scale Industries (Ghosh 2008, 15), after her visit to see HESCO's various initiatives. Recalling his early efforts to revive these watermills, Dr. Joshi recalled his childhood memories associated with Dalimshahar, a village near his native town Kotdwar where the flour used to be made by traditional *gharats*. To spread awareness about them, he started visiting surrounding villages and held a meeting on *gharat* management. Some people wanted to reject this "old technology," while others suggested other alternatives, and Joshi was pained about the overall apathy towards *gharats*. In 1991, HESCO started two experiments with their engineer colleague Damodar Dobhal and Manmohan Negi, who was trained in Najibabad. They installed ball bearings under the turbine to increase its speed and replaced wooden plates with sheets made of galvanized iron; this resulted in the increased efficiency of *gharats*. Bharat Electrical Limited (BEL) also joined in these early efforts with different kinds of turbines, such as from Intermediate Technology Development Group (ITDG) and IT Park's cross-flow turbines. But the brakes started breaking apart in Chamoli district, which took a couple of months to repair. Eventually, they simply refurbished the traditional *gharats* with turbines that were durable, sustainable, and affordable for the Himalayan villages, proving the words of an 18th-century British administrator Alexander Walker, "The practice of watering and irrigation is not peculiar to the husbandry of India, but it has probably been carried there to a greater extent, and more laborious ingenuity displayed in it than in any other country." Other manufacturers of Kumaon and Garhwal also used *panchakki* (watermill) in the crushing of ore according to other British administrators J.D. Herbert and J. Manson, who mentioned, "In reducing the ore to fragments, the Dhunpoor miners employ the panchakki or watermill. When water is present no better plan can be devised" (Dharampal 1971).

In 1993–94, an organization for all the *gharat* operators was founded, and foot-marches were conducted for spreading awareness. *Gharat* owners actively participated and accepted *gharats* as a viable and profitable technology. As of 2012, *gharats* were an active industry in Jammu and Kashmir, Uttarakhand, and other northeastern states of India. Dr. Joshi mentioned,

> Hindu texts remind us that unity is strength in the present Dark Age, i.e., *Sanghe Shakti Kaliyuge*. Today, Himalayan *gharat* owners are united and fight together for their issues and government is bound to listen to them. We are happy that *gharat* owners have awakened.

According to Joshi, there are four prerequisites for any technology in the rural context. It should be simple, cost-effective, easy to learn, and easy to maintain. As a counter-example, the International Development Research Center (IIDC) of Canada installed a watermill in the Chamoli district of

Uttarakhand, but it soon broke and its repair proved to be very expensive and time consuming. Recalling this experience, he argued,

> Until efficient technologies are universalized and decentralized, it is of little use for Indian situations. *Gharat* is a fifteenth century efficient technology for Indian villages. There are about 4,000 of them that are installed by HESCO out of total 10,000 in Himalayan villages. These HESCO-upgraded *gharats* have helped restore the dignity of *gharat* owners. An international watermill convention should be organized in which watermill operators from such countries as Switzerland, China, Iran, and Iraq should be invited. This is one of the best examples of decentralized power generation with ample scope for employment generation and electrification for thousands of villages. Watermill is much more than a mill. It is a living example of a philosophy to use local resources based on local innovation. This philosophy should be applied to many other such traditional resources.

Gharat workers now regularly approach UREDA (Uttaranchal Renewable Energy Development Agency) for upgrading their watermills, after Dr. Joshi's initial intervention. Gharat Development Council (*Gharat Vikas Samiti*) was established on June 21, 1995. I also met Matiyalji, to whom Dr. Joshi had donated 50,000 rupees from his award money, for the construction of his *gharat* center. Unfortunately, in 2011, a cloudburst destroyed his center and the government provided the relief and compensation for him, again with the HESCO's intervention. Today, Matiyalji's *gharat* is extremely busy helping several families in the neighborhood. He was proud to share with me that the then Indian Minister for Small Scale Industries Vasundhara Raje had flown to visit his *gharat* center in 1998. According to the Indian Army's website,[22]

> Operation Sadbhavana (Op Sadbhavna) was launched by the army in 1998 in remote areas of Jammu and Kashmir where terrorists and anti national elements had wrought havoc by large scale destruction of government property and public assets like schools, bridges, electricity supply system etc. causing severe hardships to locals.

The army, upon hearing about HESCO's work in Uttarakhand, invited Dr. Joshi to do similar projects in Kashmir as well.[23] Dr. Joshi took HESCO's watermill expert, Manmohan Negi, to work with the army personnel. Negi and his other HESCO colleagues were sent to Indo-Pak border in Rajouri-Poonch and installed about 1,500 watermills there. While upgrading one of the watermills, they faced the threat of terrorists and were barely rescued by the army. They encountered a similar situation in the northeastern Indian state of Manipur when surrounded by extremists. In the northeastern Indian states of Meghalaya, Manipur, and Nagaland, the Indian Army's Assam Rifles had also joined hands with HESCO in 2007 to develop Rural Technology Hubs to uplift the tribal communities.[24] These infrastructure units trained the local people, especially women, on post-harvesting technologies, bakery, candle making, horticulture, incense and charcoal making, and multiple applications of watermills.

As mentioned previously, Indian minister Vasundhara Raje was impressed with HESCO's projects and invited Dr. Joshi and his team to her political constituency Jhalawar. Recalling his visit there, Joshi told me,

> I told them that in Rajasthan, Kota's textiles and Bikaner's snacks are very famous. Similarly, Jhalawar should create its own identity based on its unique local resources such as oranges and neem. This suggestion ignited them and several industries eventually started there. This kind of model will decentralize the economy and preserve the ecology and people will be connected with their local resources. People and villages can become self-dependent with this thought.

Reviving traditional farming and food patterns

People of Himalayan regions lack proper nutrition (Pant 2004). This was abetted by the "Green Revolution," which highlighted only two crops – rice and wheat, replacing several traditional nutritious crops (Gopalan and Aeri 2001). Since 2002, HESCO started working on the revival of traditional crops such as finger millet (*mandwa*), green amaranth (*chaulai*), and lysine, which are rich in calcium and protein.[25] Farmers in about thirty-four villages are growing them with HESCO's training in collaboration with International Development Research Centre (IDRC), the M. S. Swaminathan Research Foundation (MSSRF), the University of Agricultural Sciences (UAS) at Dharwad, and McGill University.[26] Machines for reducing drudgery, such as thrashers, dehuskers, and destoners, are also provided to farmers. As part of this project, Dr. Joshi and his colleagues Dr. Rakesh Kumar and Dr. Kiran (Rawat) Negi visited McGill University for more discussions, and their collaborators have visited Uttarakhand farms to see the flowering of crops.

Recharging traditional water sources[27]

In 2003, scientists from the Bhabha Atomic Research Center, at the initiative of Dr. M. A. Chidambaram, Principal Scientific Adviser to the Government of India, visited Uttarakhand to study the severe water crisis, which led to a joint project in Rudraprayag district.[28] According to a BARC publication (Gantayet 2007),

> In the mountainous region of Uttaranchal, springs are the only available source of water for domestic and agricultural use. Insufficient spring water due to low discharge causes a lot of hardship to the people in summer. At the request of HESCO Uttarakhand, environmental isotope investigation was carried out, along with hydrogeology and hydrochemistry, to identify the recharge zone of the springs in the mountainous Himalayan region. Stable isotope (^{2}H & ^{18}O) data of springs collected from 3 different valleys indicated that the springs located in valleys 1 and 2 were getting recharged from the same area while the springs located in the valley 3 was getting recharged from a different area. Local precipitation is the main source of recharge. Hydro-chemical results indicated that the quality of spring waters was fresh and contained Mg-Ca-$HCO3$. The investigation for identification of the recharge zone of the drying springs in Gaucher Area has been completed. Based on these studies artificial recharge structures are being planned by HESCO.

During the monsoons, scientists of BARC and HESCO started collecting water samples from various catchment areas from there water discharges. According to a news report,[29] Using the environmental isotope techniques, described above, scientists first traced the recharge areas of the water sources. Using the latest isotope hydrogeochemical techniques, the scientists worked tirelessly to track down the origin of the dried springs on the slopes of the hills. Once the origin and route were traced, scientists built dams so that water would percolate within the earth. After firmly establishing the catchment area of each spring, thirty-three water dams and tanks were built with the help of locals in the recharge zones to hold the rainwater. The springs were recharged at the village downstream. After monitoring the discharge from these sources, it was found that the water level had increased considerably, and two new water springs had also emerged. As of June 2012, the water discharge in the dried spring had risen to 16 liters per minute from a scanty 2 liters per minute couple of years ago. During the peak summer season when the temperatures in the hills rise to 40 degrees Celsius, the water discharge in these springs was 6 liters per minute.

Vinod Khati, a HESCO geologist, worked with BARC for two years. He prepared a concept note based on firsthand observation of springs and streams at Gwar Chauki. Pre-monsoon and post-monsoon samples (of both rainwater and spring water) were taken, and then the zone was identified for recharging. The entire process took about two and a half years at Nagrasu and was occasionally observed by the experts from the Indian Institute of Technology at Roorkee and at Kanpur. The residence period of rainwater (the time taken to reach the spring) is analyzed using isotopes and the catchment area is kept far from the spring. Due to construction, deforestation, quakes, and cemented roads, the percolation of water is affected. As a remedy for this problem, HESCO constructed trenches, ponds, and check dams. This helped in harvesting the rainwater, which could slowly percolate and recharge the water level in the catchment area. Results were also published in scientific journals (Shivanna et al. 2008). As of June 2012, about 120 springs of Uttarakhand, Himachal Pradesh, and Jammu and Kashmir were being targeted by HESCO. Material (cement) and skilled labor cost were provided by HESCO, while nearby villages provided the stone and unskilled labor. Harmful elements, such as iron, fluoride, arsenic, and lead, were checked and water containing those elements was not recharged. The Indian Army's Garhwal Regimental Centre at Lansdowne in Pauri Garhwal has also selected a village for applying the isotope technology for recharging a water spring there. In April 2013, BARC inaugurated an isotope water-testing laboratory at HESCO's campus near Dehradun, and equipment worth more than 30 million rupees, including a mass photo spectrometer, was installed in the laboratory. BARC has already trained ten groups of youth from the hills in isotope technology for the benefit of their communities for underground water recharging. The BARC scientists using isotope hydrology technology were able to recharge sixteen water sources that had virtually gone dry, leading to a severe water crisis in the Gauchar area of Rudraprayag district of Uttarakhand.

In Nahan (a town in Himachal Pradesh, which we visited above), HESCO made the catchment area (gabion check-dam structure; made of wires and stones) and a percolate pond, with trenches in between. In my fieldwork, I observed that the rainwater was stored in the gabion structure and a percolation pond recharged the spring. About fifteen such structures were made within an eight square kilometer area in nine neighboring villages with a total population of 2,500. The principal scientific advisor of the central government of India had provided funding for the mason and stonework, while unskilled labor came from the villages, as is the usual HESCO model. Villagers also submitted a resolution to the local forest department to obtain permission to construct these structures. In the converged land, owned by the Public Works, Revenue, and Forest departments, around the structures, trees were planted. Also following the HESCO's integrated approach to sustainable development, separate groups of women, artisans, and bamboo workers were trained to generate employment.

Moving on, once in 2005, while on his pilgrimage to Binsar Mahadev in Pauri district, Dr. Joshi encountered a man who was about to sell his daughter named Bansuri. Upon Joshi's intervention, he challenged Joshi to adopt her and Joshi instantly accepted. Ever since, Bansuri has lived with Joshi's family. As of June 2012 when I met her, Bansuri was completing her undergraduate degree in music and Joshi was planning for her marriage after her graduation. Similarly, Joshi had earlier sheltered a child named Manmohan Negi and had sponsored complete treatment of his knee. As of June 2012, Negi was spearheading the development and installation of *gharats* in hundreds of Himalayan villages, as noted previously.

Prasad project: connecting the local ecosystem with pilgrimage centers

In 2004, Joshi was given the Sat Paul Mittal Award, named after one of the telecom tycoons of India Sunil Mittal's father who was a parliamentarian from Punjab in the 1970s and 1980s. The year also marked the launch of the Prasad Project for Hindu pilgrimage centers, such as Badrinath, Kedarnath, and Vaishno Devi, in collaboration with state governments. Raw materials are kept ready around October every year, and people in nearby villages prepare dessert (*laddoo*) boxes according to the training received from HESCO. Every year, major Himalayan temples open after winter, and about 500 kilograms (900 pounds) of *laddoos* are available for sale outside these temples to the pilgrims, with special care taken for fragile or perishable items. *Ringal* baskets and incense sticks are also designed and sold there. These products are advertised on local TV channels during the pilgrimage season in the summer. Recalling the launch of this project, Kiran (Rawat) Negi told me,

> Although the temple boards were resistant about these products in the beginning, they allowed them after local women's activism supported by us. We approached both Hindus and Muslim places of worship to promote the sale of local products. We preferred to use the secular language because it touches every class and section of the society instead of giving preferential treatment to any one section. These products help generate employment for local people and are mainly run by the women. There are two groups for designing the baskets and three for making *laddoos* with about twenty people in each group. In the first year, we had a total of sixty such groups at different temple areas. We had a sale of 120,000 rupees from 250 kilograms of *laddoos* in just ten days in 2011. As of June this year, we have already earned 15,000 rupees. In ancient times, these were the *Prasad* (offerings to the deity

made from local products by the local people) at the temple but recently cardamom and other outside materials are being preferred which has resulted in the loss of local employment. We are just trying to revive the traditional economy based on the local ecosystems. Recently, at the Garuda Ganga temple, we have started selling our products there, which are being appreciated and we are getting very good orders. We have also revived the traditional coal-based incense sticks instead of the outside incense sticks sold in markets.

Continuing the HESCO chronological developments, in 2005, Joshi was given the TN Khoshoo Memorial Award (named after a renowned environmental scientist and administrator) for his work to promote sustainable livelihoods in the Himalayas.[30] In 2005, *Harmony* magazine honored Joshi as one of its "Silvers of the Year."[31]Also in 2005, two high-altitude villages near Kedarnath, Kali Math and Okhi Math, suffered a major landslide, and the affected villagers had climbed up the hill named *Kaal Shila.* The HESCO volunteers had to climb up the mountains to help the victims with food and medicines and also to recover several dead bodies.

According to a news report[32] (and confirmed by my fieldwork visits and interviews), at the Indian Independence Day Celebrations on August 15, 2005, The Doon School, India's most prestigious high school located in Dehradun, invited Dr. Joshi as their Chief Guest. He asked the school to help in his mission of making villages self-sufficient. It was then decided that the school and HESCO would work together in Fatehgram in Vikasnagar Tehsil (about 18 miles from Dehradun), a village of sixty-five people of nine families, which had no road, electricity, or potable drinking water. The students then got down to doing their homework and came up with a step-by-step plan of action. Students went to the village on weekends and vacations and after an initial survey, prioritized the needs of the villagers. Electricity was decided on first, so work on upgrading two *gharats* began. Earlier, these watermills were only used for grinding wheat and cereals. With the installation of new turbines, the *gharats* started generating three kilowatts of power each, enough to light up homes of all nine families. The efficiency of watermills also increased and their owners could grind more wheat, resulting in better incomes. None of the houses had toilets, so the students drew up a plan and a toilet was built inside each house. A school building, a half-kilometer approach road to the village, and a 400-meter irrigation channel were built. Each family living below the poverty line, in consultation with students and HESCO, were provided several income-generating activities. One of the families wanted a fishpond, while another wanted to start a plant nursery. Students pitched in for both ventures, helping them out while the school provided funds. One of the families was trained in beekeeping and was provided with two boxes of bees. The HESCO team and students taught another family to make cement blocks. Soon, rooms replaced the two dilapidated watermill huts. Doon School started this project by investing 150,000 rupees, and the project continued until May 2011, as told to me by Dr. Mohan Joshi, the Dean of Activities and teacher of Hindi and Sanskrit, whom I met during my tour of the Doon School in June 2012.

In 2006, Joshi led a bicycle rally from Jammu to Dehradun, glimpses of which we saw above in Dwarika Semwal's memoirs. Also in that year, HESCO found its current office space (*Ashram*) and shifted to Shuklapur from Mehuwala. At the Ashram, a new lantana cottage was designed for Joshi, and he shifted there on his birthday[33] on April 6. The land was bought for about 280,000 rupees; the money came from the savings from various awards received by Joshi. In my visit, I noticed several models for different rural technologies that are displayed here, reminding me of these words of a noted Gandhian thinker Dharampal (2000):

> The National Council of Science Museum, and the Indian state, which will ultimately have to finance any such nation-wide plan, should consider the early setting up of technological museums; if not in every taluk, at least in every district of India. While such museums will also display products of modern western science and technology, their main display should be of indigenous artifacts from the respective area, or the region surrounding it.

According to Dr. Joshi, "We can provide models and conduct experiments but we cannot become the system." He also told me about a small river coming from the nearby Asharodi Range, whose flow has increased to 600 liters per second by HESCO's recharging and weekly cleaning efforts, in which I too participated once. As one of the news report mentioned (and I saw in person at the HESCO Ashram in Shuklapur),[34] Dr. Joshi is interested in harmonizing the four elements of *Chinta* (planning and policy), *Chintan* (discussion with the society), *Chaitanya* (self-consciousness), and *Chita* (pyre). According to Joshi, nobody dies; dead was the one who did not know how to live. Pyre was the beginning of the second life for any person who died, and was brought to the HESCO Ashram for cremation. He showed me a plant and told me that the person cremated here years ago was still alive in that plant. On the one side was the Shiva temple and, at a little distance from it, was the place where the bodies were consigned to fire, marking the beginning of another life in the form of a sapling. Once the body is cremated, its ash is put into the pit dug for the sapling. As Joshi mentioned, "We all are made of various elements and material and when the ash of the body is put into the pit it provides the sapling with minerals and salts which help in its growth." As of June 2012, Joshi began his day at the Ashram temple offering prayers to Lord Shiva, often called the "Hindu god of destruction." But, at this place where Shiva was being worshipped, a lot of creativity was also going on with the synthesis of dharma and science. As Dr. Joshi argued, the malady was in the very idea that science is different from dharma, and things will change only when science and dharma are harmonized. "We have to rise from the selfishness and think about community and HESCO is busy doing the same," added Joshi. As exemplified by the rebirth example here, HESCO was determined to lead the campaign to harmonize science and dharma. In addition, HESCO, in conjunction with the Dehradun-based Forest Research Institute (FRI), had adopted the nearby area of around 50 acres and was developing the concept of system recharge, i.e., inclusive and sustainable development including recharging the degenerated ecosystem; rejuvenating water bodies, soil surface, grass-herbs-forest cover, and the food chain; and reducing the human–wildlife conflict and invasion of wildlife on the village resources.

In 2006, the Jamnalal Bajaj Award (named after a renowned Gandhian industrialist and philanthropist),[35] with 500,000 rupees, was given to Joshi, but as with earlier awards, he deposited this money into HESCO's account, even after persuasion to use the money for his son's education.

In 2006, Joshi was awarded the Padma Shri Award by the Indian government for social work.[36] For the award selection process, when the local income tax department investigated his tax returns, they were so impressed that they also honored him for living a simple and charitable life, as revealed to me by Dr. Rakesh Kumar.

In October 2006, Dr. Joshi was invited to deliver the lecture "Power to the People: Producing Electricity with Watermills for Poverty Alleviation" by the Technology and Sustainable Development Group at the University of Twente in Enschede, Netherlands.

In 2007, the Jindal group offered one million rupees to HESCO for rehabilitating eighteen villages in the wake of the construction of their steel plant in Odisha. However, HESCO accepted only the reimbursement for boarding, lodging, and traveling, denying the rest of the cash amount. According to Dr. Rakesh Kumar, HESCO wanted to remain voluntary, not a usual NGO whose agenda is driven by money. Dr. Kiran (Rawat) Negi, Dr. Rakesh Kumar, and Prem Khandwal worked in the village Jahajpur near Puri in Odisha and taught bakery technology and horticulture skills there. Savitri Jindal, the owner of the OP Jindal Group, came to HESCO in her personal helicopter to thank them for this consultancy.

Dr. Joshi participated in the 3rd Global Forum on "Hydropower for Today" in June 2007 in Hangzhou, China.[37]

In 2007, Ritwick Dutta, a Supreme Court lawyer, conducted a workshop on RTI (Right to Information Act) with HESCO. This law is widely seen as an empowering tool available to citizens to request any information from government offices, making them more transparent and accountable. In 2008, Joshi led another cycle rally from Kanyakumari to Dehradun, which was launched on January 12 on the birthday of Swami Vivekananda. As of December 2012, he has led about six foot-march and seven cycle rallies. Also in 2008, the media giant CNN and its Indian partner IBN, in association with Reliance Industries, gave the Real Heroes Award to Joshi for rural development.[38]

In 2008 (and earlier in 2001), HESCO was given the National Award for women's development through the application of science and technology.[39]

In 2008, the Project, "People and Protected Areas: Conservation and Sustainable livelihoods in partnership with local communities" was a joint venture by the Science for Equity, Empowerment & Development (SEED) Division of Department of Science and Technology (DST), Government of India and WWF-India. It aimed to encourage the thirteen NGOs developing new techniques to improve economic status of people living near the Protected Areas. The team came for a one-day exposure visit to HESCO, Dehradun on June 12th 2009. In 2010, India's foremost English newspaper The Times of India honored Joshi with the award "Heroes of Hope".[40]

In 2010, India's foremost English newspaper, *The Times of India*, honored Joshi with the award "Heroes of Hope."[41]

During my fieldwork, I met Kumari Bhagirathi Panwar in Athala, a Scheduled Tribe-dominated village. She is a carpet weaver, and in 2010, HESCO helped her with a water tank, steel filter for drinking water, and a vermicompost bin for her kitchen garden. She was given a new frame, loom, and one-month training for carpet weaving and designing. Other folks in the village were also provided training for fruit processing, creating a fodder bank, and biscuit making and were given sewing machines. Swati Mitra from the National Institute of Design at Ahmedabad was the adviser for this training workshop and was funded by CAPART. First, a survey was conducted and discussions were held with the villagers. Farmers also provided training, which came from Saldhar where HESCO had conducted a similar training in 1991. HESCO's goal is to transfer expertise to the villages instead of making them dependent on outside "experts." The traditionally designed carpets here are sold within the Himalayas. An average carpet of 6 feet length and 4 feet width sells for 3,500 rupees, and a sofa cover sells for 2,500 rupees; mostly bought by the tourists or sold in cities as far as Kolkata or Delhi. As Panwar told me,

Overall, HESCO's interventions improved the quality of our traditional skills instead of the need to learn a new skill or technology.

Christian missionaries had also approached us, and some of us (Scheduled Tribes and physically disabled people) received professional training at Bijnaur, Najibabad, and Badapur. But, our real long-term benefit has happened because of HESCO's committed people who work with us for years together and are always here for further help.

Another village that I visited was Salur in the Chamoli district, where HESCO started working in 2010. This village is at an altitude of about 7,000 feet, with harsh living conditions especially in the winter. Its nearest town Joshimath is about six miles away, and in the absence of proper transportation, people still walk to the town (as of June 2012). Their day starts with cleaning after their cattle and then going to the forests for fodder collection and finally working in their farms. Most men have migrated to cities in search of employment and women do most of the work in their villages. There are forty-seven families here. Sunitadevi Negi is the head, *Gram Pradhan*, of the village. Most women here have completed their middle school education (until eighth grade). HESCO's Lakshmi and her husband Shersingh Rawat have been conducting door-to-door training here, e.g., fruit-processing training is done depending on the availability of seasonal fruits. Training was conducted for the whole village with the first meeting with its head, *Gram Pradhan*, and then scheduling time for a couple of hours at their convenience. About three- to five-day training camps were conducted for fruit preservation, processing, and beekeeping, in which people of all castes participated together.[42] A flourmill was already installed and fruit-processing units (juicer, pulper, etc.) were forthcoming, as of June 2012. The villagers took care of the boarding and lodging for the HESCO trainers. About one hundred such villages were approached by HESCO, from Karnaprayag to the Joshimath area near Badrinath. Farmers were involved in growing green and red amaranth, red kidney beans, and millet, whose seeds were taken from research institutes in Almora or Pantnagar. Apple plants were brought from Himachal Pradesh and distributed to encourage horticulture. As of June 2012, different varieties of seeds of wheat, mustard, vegetables, and kiwi plants were being planned for distribution in Salur. Only traditional (non-hybrid and non-genetically-modified) seeds were used, so that they could be stored for future use, obviating the need to buy the seeds every year and avoiding the use of chemical fertilizers or pesticides. After their crops are harvested, all the seeds are returned back to HESCO to be given to other villages. Seeds are continuously distributed, maintaining diversity among farmers. Some farmers are given ginger, some turmeric, and others arabi, etc. Similarly, some are given tomatoes and others okra. Local HESCO coordinators, based on the participation in meetings and workshops, decide the first batch of beneficiaries for the distribution of seeds, plants, and machines, etc. The end products are sometimes gifted back to HESCO by villagers, which are shared among HESCO members who in turn take different kinds of gifts to villagers to establish and maintain relationships. Recollecting their memories, women of Salur told me,

> Shersinghji and Lakshmi Devi gave us *chaulai* seeds of 144 white, red, yellow, pink, and green varieties. Initially, we did not trust them because in the past many NGOs came, conducted meetings, and disappeared. But, with HESCO's continuous help, we have had a very fruitful and profitable crop, about two to three times more. They also bought our crop on very good rates. We also make our breads (*rotis*) and desserts (*laddoos*) from *chaulai*. They also took us for training to Dehradun and Pantnagar and we were trained and funded to make vermicompost pits. We don't use chemical fertilizer anymore. We have also been given red kidney bean seeds (*rajma*) this year. We have also learned making millet biscuits and snacks. We consume them and also sell them. We most welcome HESCO brothers and sisters who have helped changed our lives otherwise we would have wasted our lives without proper knowledge. We will try to add more families in coming years for distribution of seeds.

Continuing HESCO's chronology, in 2011, the Construction Industry Development Council (CIDC) awarded Dr. Joshi with the Vishwakarma Award (named after the Hindu god for design and construction) for Social Responsibility.[43] CIDC is a joint venture of the Planning Commission of Government of India and the Indian construction industry. It is an umbrella organization for the construction industry in India and has a national presence with branches across the country.

In May 2011, Dr. Joshi delivered a lecture at the University of North Texas (UNT)–hosted conference of the Society for Technology and Philosophy[44] (where I met him for the first time, as alluded to in an earlier chapter). In September 2011, the Integrated Centre for International Mountain Development (ICMOD) organized the Green Economy Conference at Kathmandu, Nepal, and Dr. Joshi was one of the participants there.[45] In October 2012, Dr. Joshi revisited McGill University and UNT and delivered several lectures on Himalayan ecology and economy.

Within India, Dr. Joshi continues to address at national conventions of scholars and coordinators of different NGOs, resulting in several bridges built among NGOs and research institutes for seed distribution and other technological support. As an example, in 2012, the Indian Veterinary

Research Institute (IVRI), at Bareilly, UP, organized a camp at HESCO, and their four scientists treated about four thousand domestic animals without charging anything to the farmers.[46]

On August 4, 2012, nearly twenty-eight people were killed and hundreds were rendered homeless following heavy flooding of Assi Ganga River in Uttarakashi, where a series of hydropower projects are being developed. Although Dr. Joshi blamed such projects for the devastations,[47] he argued,

We should develop alternatives instead of only protesting the big dams (*Vikalp* instead of *Virodh*). Even a minute's power loss is intolerable to all such protesters, activists, including the religious and social leaders; they want all the luxuries but oppose the hydropower projects and dams. One side is pro-dam and the other is anti-dam. The anti-dam folks have not suggested the alternative energy sources that are needed for their luxury. HESCO advocates the middle path between the two extremes. HESCO tapped the local resources and tributaries instead of interfering with the main river. Such alternatives are the answers, the middle path, between the two extremes. All the local resources should be harnessed instead of the big dams. Smaller power plants, rainwater harvesting, and *gharats* are preferable and more sustainable. Such alternatives support both the environment and the development. First, they (policymakers and corporates) destroyed the ecology and now they (so-called environmentalists) want to destroy the economy by disrupting the big dam projects after they are almost complete. The protest should be done before the construction has begun instead of waking up so late! Before the construction of any dam, local people must be consulted and included in the discussions. The rehabilitation must also be done before the launch of the project. When I participated in a campaign on water issues, I realized that the government would not be able to provide water and electricity because there is an acute scarcity of both. Today, big industrialists are doing "contract farming" and selling their produce to Walmart and other big stores. They should similarly harvest their own water that they sell in bottles instead of stealing it from our rivers, springs, ponds, and lakes. We should raise all such issues at the election time before we cast our vote. For the first time in 2010,[48] we created the public manifesto and gave it to all political parties. Politicians are accountable to the society, not to their own parties. Local political issues must be based on local resources. Instead of industry-based economy and politics, resource-based politics and economy are needed. When our villagers will be socio-economically empowered, they will also control their political destinies.

During my fieldwork, I joined HESCO coordinator Dr. Kiran (Rawat) Negi to a village near Udaipur in Rajasthan. Here, a majority of the people were of a Scheduled Tribe engaged in subsistence agriculture of crops such as wheat, mustard, maize, and black lentils, which depend on rainwater, and vegetables, which use underground water. Fruits naturally available in the surrounding forest were mango, timru, jambul (*jamun*), custard apple (*sitafal*), Indian gooseberry (*amla*), papaya, date, and Carissa caranda (*karonda*). Mahua tree is also used and conserved for making liquor (from its flowers) and extracting oil (from its seeds). Negi addressed the villagers and suggested value addition for fruits in the form of making pickles, jams, biscuits, toffees, squashes, and chips (from potato). She suggested that by creating such local small-scale industries, youth can work in the village for themselves instead of going out to work for factories. Another botanist, Dr. S.C. Chaudhary, suggested many organic methods to make fertilizers and pesticides (by using cow urine, cow dung, jaggery, fish-water, milk-water, neem-leaves, and garlic). Rohit Jain, a native of this village, later visited HESCO's Ashram in Dehradun in January 2013 and explored the possibility of starting a fruit-processing unit at his village. In March 2013, a couple more people received training for a bakery. Projects on lantana and millet were also planned.

As of June 2013, HESCO was collaborating with about forty NGOs in diverse fields, with its continued focus on applied and participatory research with hands-on training in the farms and fields. In almost all the HESCO projects, the one universal theme is to combine traditional wisdom with newer ideas. HESCO's theory and methods are about refurbishing and renovating older ideas instead of replacing them. Some of its collaborators are SURE (Society to Uplift Rural Economy at Barmer, Rajasthan), Madhya Pradesh Science Assembly (*Vigyan Sabha*) at Bhopal, MP, Himalayan Research Group at Chhota Shimla, HP, and Bethany Society at Shillong, Meghalaya. Once in every five years, a national meeting is held of all such partners, and HESCO coordinators continually visit these different NGOs for monitoring and learning. According to Dr. Rakesh Kumar, in the initial decades of the 1950s and 1960s, the major funding from the Department of Science and Technology was being provided to the South Indian states, but with Dr. Joshi's intervention, DST formed a separate committee for Northern India, paving the way for more projects and funding for the North Indian states. Commenting on his own philosophy about social work and highlighting that veteran environmentalist Sunderlal Bahuguna was living with him at the HESCO Ashram near Dehradun, he said,

> Social workers should live at social places instead of private homes – so that they are always immersed in the society. The current social workers should also care for aged social workers that dedicate their entire lives for society even while ignoring their own families and children. When I see the past, present, and future generations and Sunderlal Bahuguna, they remind me of my own old age motivating me to focus my energies for more constructive work. Social workers should sacrifice their personal interests. Social workers should be beyond any limited causes or issues. Once an economist in Delhi asked me about HESCO's mission. Our mission is not carved in stone and adapts to the needs of the society. I am lost in a huge sea with water all around me. I am fighting in all directions to remain afloat. I am happy to have reached at least few destinations along the way. My duty was to ignite the issues and then leave it to the people to take their issues forward. My own journey continued in different directions instead of sticking to a single issue. We created the models and advised the system (policymakers) to incorporate them. We are unable to create the new system but I am happy that the governments have included our suggestions in their development agenda such as lantana, fruit processing, and *gharat*. In some ways, we are ahead of the Buddha, Vivekananda, and Gandhi, who had not experienced the life of farmers in Indian villages. With our experiments, we succeeded or failed but at least we attempted. Our future mandate is open and we will raise issues on all fronts. We are thinking of raising the issues of plastic consumption, rainwater harvesting, and creating bicycle lanes in the cities such as Dehradun. To serve the society is the best way to reach God because you become a replica of God in service and fulfill all your dharmas.

However, during my fieldwork in some of the villages, I was a bit surprised to see the emphasis on revenue generation and raised the concern with Dr. Joshi. His response was,

> Yes, villagers are being urbanized and industrialized at a very small scale by our work but this is different from the real cities and big industries. In the cities, people do not produce the raw material and they do not depend on their physical labor to produce raw materials. In the villages, people do not have the luxury to divert their energies into other leisure activities unlike urban people. In the cities, people get their finished products directly without worrying about the raw materials. A village is

still a social community unlike a city where neighbors remain unknown to each other. This rural social connectivity deters people from doing anti-social actions. Villages have been living in extreme poverty for many centuries so initially wealth may encourage them to run after few luxuries. But why should cities control what luxuries villagers can use? The same standards should be applied to cities and villages. If villagers are earning their luxuries by their real hard physical labor, there is nothing wrong in it. This will encourage them to continue doing farming and other rural work. To be completely dependent on luxuries is wrong. Wherever HESCO has reached, people honor us for our simplicity and sacrifices and eventually they will also follow our examples. So, no need to force them to renounce new found luxuries at this time. Let them experience the pleasures and let them decide the upper limit of their wealth. Renunciation is successful only for those who have experienced the pleasures and have realized that the resources are for you; you are not for resources. It is the urban people who form the opinion of any issue in the world today. Most awards are given to NGOs working in urban areas that have little grassroot efforts. As long as one's back is towards the city, you are connected with basic issues. But if one's back is towards the village, one will forget the basic issues and start running after materialism. HESCO chooses the middle path combining the Gandhian philosophy of self-reliance with modern technologies.

Notes

1 Under the firms and societies Registration Act (1960) in 1983 and under FCRA in the year 1991 to ease the formalities required by many associating agencies.
2 However, the gasifiers are no longer used after the arrival of electricity facilitated by the government electricity board.
3 http://www.scienceandsociety-dst.org/coresupport1.htm (accessed in June 2013).
4 http://www.downtoearth.org.in/node/23584 (accessed in June 2013).
5 http://www.governancenow.com/gov-next/green-gov/ngo-markets-corbett-furniture-provide-jobs-villagers (accessed in June 2013).
6 http://www.indianexpress.com/news/kisan-bank-a--kind--way-to-lend-cash/416707/ (accessed in June 2013).
7 http://www.dnaindia.com/india/report-laurie-baker-remembered-in-uttara khand-1088409 (accessed in June 2013).
8 http://www.downtoearth.org.in/node/16209 (accessed in June 2013).
9 Another foreigner was a woman ascetic from San Francisco, Kamla Maa, who joined some of the HESCO campaigns and wrote a book titled *Son of Himalaya* on Dr. Joshi, but Joshi suppressed its publication because "he did not want to be deified" (as told to me by Dr. Rakesh Kumar).
10 In 1991–92, trees worth millions of rupees were being poached and logged illegally. When he started protesting about it, he was given good coverage in the local news media. Once, however, a huge crowd of hooligans attacked during

a meeting where Joshi was about to give a speech. Fortunately, he was rescued with police intervention.

11 http://india.ashoka.org/fellow/anil-prakash-joshi (accessed in June 2013).

12 The harvested and stored water is used for domestic animals, while natural water in Himalayan rivulets, called *Dhara* water, is used for drinking.

13 In the winter season, heavy snow on the upper echelons of the mountains forces leopards to invade the villages in the lower altitudes and take away domestic animals.

14 However, in some villages, rainwater harvesting is no longer done after the local municipal corporations started providing water.

15 Along similar lines, Joshi helped bamboo workers, watermill operators, blacksmiths, masons, and numerous other local professionals in getting organized. Many different organizations were formed for different professions to help them raise their voice as an independent force.

16 http://www.tribuneindia.com/2009/20090103/dplus1.htm (accessed in June 2013).

17 http://www.aagaas.org/about_us.html (accessed in June 2013).

18 http://www.thehindu.com/sci-tech/energy-and-environment/wakeup-call-for-a-region-facing-himalayan-problem-ngos/article3878167.ece (accessed in June 2013).

19 http://ubfdb.org/index.php?option=com_content&view=article&id=79&Ite mid=96 (accessed in June 2013).

20 http://www.trust.org/item/20130314152900–5tbfo/?source=spotlight (accessed in June 2013); http://www.csmonitor.com/World/Making-a-difference/Change-Agent/2013/0315/Updated-water-wheels-power-India-s-rural-mountain-economy (accessed in June 2013).

21 http://articles.economictimes.indiatimes.com/2002–12–14/news/27335636_1_baba-amte-magazine-social-worker (accessed in June 2013).

22 http://www.indianarmy.gov.in/Site/FormTemplete/frmTempSimple.aspx?Mn Id=mfwtOFq27Ziy8WwX0msTGw==&ParentID=i9mGjT+eKQKJk1sfyC5 EMg== (accessed in June 2013).

23 http://mod.nic.in/samachar/nov16–04/body.html#l6 (accessed in June 2013).

24 http://news.oneindia.in/2007/07/25/assam-rifles-and-hesco-to-train-tribals-to-be-self-reliant-1185361814.html (accessed in June 2013).

25 Just as American grocery stores also carry several multi-grain food products that are more nutritious.

26 http://www.thehindu.com/todays-paper/tp-national/tp-karnataka/millet-production-to-be-promoted-in-four-districts/article951094.ece (accessed in June 2013).

27 http://www.rediff.com/news/report/barc/20061107.htm (accessed in June 2013).

28 http://www.barc.gov.in/publications/eb/golden/chemical/toc/chapter3/3_6.pdf (accessed in June 2013).

29 https://www.tribuneindia.com/2013/20130317/people.htm#3 (accessed on May 8, 2019).

30 http://www.atree.org/KhoshooAward (accessed in June 2013).

31 http://www.harmonyindia.org/hportal/VirtualPageView.jsp?page_id=6950 (accessed in June 2013).

32 http://www.indianexpress.com/news/in-a-village-doon-school-s-extended-classroom/3545/2 (accessed in May 2012).

33 Joshi's birthday is celebrated on April 9 in some villages, which he initially did not approve of.

34 http://www.thesundayindian.com/en/story/life-begins-after-death-at-hesco-ash ram/14/35356/ (accessed in July 2012).
35 http://www.jamnalalbajajfoundation.org/awards/archives/2006/science--and--technology/anil-prakash-joshi (accessed in June 2013).
36 http://mha.nic.in/pdfs/padma2006.pdf (accessed in June 2013).
37 http://www.g77.org/pgtf/finalrpt/INT-06-K10-FinalReport.pdf (accessed in June 2013).
38 http://www.realheroes.com/dr-anil-joshi.php (accessed in June 2013).
39 http://www.scienceandsociety-dst.org/NationalAward.pdf (accessed in June 2013).
40 http://articles.timesofindia.indiatimes.com/2010-01-02/india/28143167_1_arvind-kejriwal-ela-bhatt-toi (Accessed in June 2013).
41 http://articles.timesofindia.indiatimes.com/2010–01–02/india/28143167_1_arvind-kejriwal-ela-bhatt-toi (accessed in June 2013).
42 Newer styles of home in the villages need newer kind of beekeeping and the information for such bee boxes is given.
43 http://www.cidc.in/new/support/viswakarma/New/3VKA-FL_Web_28Feb11.pdf (accessed in June 2013).
44 http://philosophy.unt.edu/sites/philosophy.unt.edu/files/SPT%20Program%20Web.pdf (accessed in June 2013).
45 http://www.icimod.org/?q=3734 (accessed in June 2013).
46 http://www.ivri.nic.in/Annual_Report_IVRI.pdf (accessed in June 2013).
47 http://www.indianexpress.com/news/flooding-cloudbursts-at-uttarkashi-due-to-hydel-projects/987401/ (accessed in June 2013).
48 http://www.tribuneindia.com/2010/20100828/dun.htm#3 (accessed in June 2013).

5   Restoring a Holy River in Punjab

Having heard about the efforts of Baba Balbir Singh Seechewal (hereafter Baba), I contacted their office near Shahkot, a town near Jalandhar in Punjab, in December 2013. Later, I arrived at their office and soon was taken for a boat ride on the Kali Bein River, which was restored to its original purified state after the massive efforts of the Sikh community led by Baba (see Figure 5.1). What I present below is based on my interviews with Baba and his colleagues. In addition, I have extracted relevant information from their relevant literature that is based on the writings of Baba and his team. A 2008 master's thesis from the University of Manitoba, Canada, by Manpreet Nigah was also useful (Nigah 2008).

Figure 5.1 Baba Balbir Singh Seechewal

Balbir Singh was born on February 2, 1962, to Chanan Kaur and Chanan Singh at Seechewal, a small village near the town of Shahkot in Jalandhar district. After his elementary school education in his village, he went to government high school at Nihaluwal, and then went to DAV (Dayanand Anglo Vedic) College at Nakodar. Here he met a Sikh religious leader named Avtar Singh in 1981 and was inspired to renounce worldly life. He dropped out of college and dedicated himself to the service of his guru Avtar Singh. When Avtar Singh settled at Nirmal Kuteya in the village Seechewal, he trained Balbir Singh with the religious philosophies originally learned at Haridwar. On May 27, 1988, Avtar Singh was brutally attacked by his detractors and soon after chose Balbir Singh as his successor. After taking over the leadership, Balbir Singh continued the philanthropic activities of his guru. His early focus was on constructing the roads connecting the remote villages of Punjab, which gave him the name *Sarakon Wale Baba* (Baba of the roads). Planting saplings of flower and fruit trees and installing underground sewerage systems in the villages and towns, he earned further reputation in Punjab. Based on his own early student life filled with economic hardships, he opened a school and college in Seechewal. Finally, when he realized the plight of the Kali Bein in 2000, he plunged into action to clean the river (Nirmal Kuteya 2004).

The Kali Bein: an overview of the river

Although Punjab literally refers to the land of five rivers, after the partition of India and Pakistan in 1947, the Indian state of Punjab was left with only three rivers – Ravi, Beas, and Sutlej flew in the Indian state of Punjab. Formed by alluvial deposits of the delta of these rivers, Punjab's soil is rich and fertile, and the state leads the country in agricultural production. However, in the last fifty years, population growth, urbanization, and industrialization led to the deterioration of natural resources such as land and rivers. The so-called "green revolution" included intensive use of poisonous chemical fertilizers and pesticides that polluted the already depleted natural water sources. Especially during and after the 1950s and 1960s, traditional water resources were abandoned for the easy availability of ground water accessible to most of the farmers with free electricity from the government. According to Punjab government sources, the water table in Punjab has been declining at an average rate of 0.23 meters per year during last two decades. It now shows a fall of 24 cm to 25 cm per annum. In the Doaba region, located in the center of Punjab, the situation is even worse, as there is no widespread canal network, agriculture mainly depends on ground water, and the number of tube wells has increased more than three times since the 1970s. In the two major districts of Jalandhar and Kapurthala, extraction of water exceeds recharging by more than 50%. One of the major sources of recharging the water table is the river Kali Bein, which became the focal point of Baba Balbir Singh's efforts (Souvenir – Ek Onkar Trust 2006).

Kali Bein literally means "black stream" in Punjabi, as *Bein* seems to have been derived from *Veni*, which means "stream" in Sanskrit. The term *Veni* is commonly used for river such as *Triveni*, which refers to the conflu-ence of three rivers of Ganges, Yamuna, and the mythical Sarasvati at Alla-habad. *Veni* also refers to women's braided tress of hair, in several Indian languages. Kali Bein thus refers to the black stream flowing in the plains of Punjab. It is named *Kali* (black) because of the water's hue (Souvenir – Ek Onkar Trust 2006).

The catchment area of the Kali Bein is 945 square miles, length 90 miles, and the average slope is 13.5 foot per mile. It rises from the Terkiana marsh-land in Mukerian village of Hoshiarpur district. Although in ancient times it was a tributary to the river Beas, over time Beas changed its course and the Kali Bein started taking its water from the waterlogged areas and marshes of Terkiana, where underground water oozes out from several villages such as Dhanoa, Himmatpura, Vadhaia, Shatabkot, and Terkiana. Beginning its jour-ney from Terkiana marshland in Mukerian village of Hoshiarpur, the Kali Bein moves in a southwesterly direction, coiling like a snake (Souvenir – Ek Onkar Trust 2006).

Together with other rivers, the Kali Bein is appropriately called the life-line of Doaba, the central region of Punjab. The Kali Bein drains out the excess water of the waterlogged Bet area of Mukerian and Dasuya villages in Hoshiarpur district and brings it to a large area of Kapurthala district, where it is much needed for the recharging of water table. During the mon-soon flood times, the Kali Bein comes to rescue the people and their prop-erty by draining out the floodwater. It is thus vitally important for the Doaba region of Punjab in several ways (Souvenir – Ek Onkar Trust 2006).

The Kali Bein is especially significant for the history of the Sikh religion. Around 1499 CE, Guru Nanak, who later emerged as the founder of the Sikh religion, came to stay at Sultanpur Lodhi, on the banks of the Kali Bein, where his elder sister Bebe Nanaki was married to Jai Ram, a revenue officer of Daulat Khan Lodhi, the Nawab of Sultanpur Lodhi. Nanak was employed as an accountant by the Nawab. Being spiritually inclined, he used to meditate regularly. Once he went to take his daily bath at the Kali Bein and disappeared inside the water. On the third day, when he emerged out of water, he recited these words, which later became the opening verses of the Guru Granth Sahib, the Sikh holy text: "One universal Lord, true name, creator, all-pervading, no fear, no animosity, timeless being, beyond birth, self-existent, by the guru's grace." This idea of oneness of God breathed a fresh air into the Indic religious traditions, and a new humanistic philosophy was thus born on the banks of the Kali Bein, which later became the Sikh religion. Sultanpur Lodhi now has a grand Sikh Gurudwara at the site of Guru Nanak's meditation, which is regularly visited by thousands of Sikhs and others. The major Sikh holy places along the banks of the Kali Bein include Gurudwara Sant Ghat (the site of Guru Nanak's enlightenment), Gurudwara Ber Sahib (the place where Guru Nanak used to take his daily bath and where he planted a berry tree which still survives today), Gurudwara Pul Pukhta (related to the sixth Sikh guru Hargobind), and Gurudwara Rababsar at Bharoana (from where Guru Nanak's companion Mardana got his musical instrument, *rebab*, to play with Guru's spiritual recitals). People living near the Kali Bein have a long tradition of taking a holy dip in its water, especially on the special occasions of Vaisakhi, full moon night, and new moon night. Burial grounds of many villages are also located around the Kali Bein. Emotional and religious relationship of people with the river has produced much folklore, poetry, and paintings. Especially noteworthy are books based on Guru Nanak's biography such as *Vishav Noor* by Avtar Singh Azad, *Param Purakh* by Inderjit Singh Tulsi, and *Nanakayan* by Mohan Singh (Souvenir – Ek Onkar Trust 2006).

The problems of the Kali Bein: pollution, encroachment, and scarcity of water

Like many other Indian rivers, the Kali Bein has lost its earlier glory and purity. It fell prey to the commercialization, urbanization, and industrialization of the Doaba region of Punjab. The government and business leaders were busy celebrating the economic growth while their environmental ignorance brought havoc to the river. Most of the villages and towns on the banks of the Kali Bein freely drained their sewage into it. The sites allotted for making ponds to serve as storage for sewage waters vanished due to unlawful possessions by local business leaders. In many villages, the government even sanctioned grants for such projects to install sewerage systems to let the drainage go directly into the river. Such actions have led to the environmental problems in at least three different ways (Nirmal Kuteya 2004).

First, the polluted effluents, including poisonous chemical waste of factories and highly infectious waste of hospitals, pollute both the water of the Kali Bein and the ground water around the Kali Bein, which leads to polluting the drinking water and thus causing long-term damage to public health. Already several fatal diseases, including cancer, are spreading in the region with a so-called "cancer express" train that patients regularly take to Bikaner, Rajasthan, for treatment (as mentioned in several media reports such as one by *Indian Express* on December 1, 2013). Second, the polluted sewage water flowing into the Kali Bein brings clay mud with it and spreads it over the bed of flowing water to make a thick layer of silt, thus clogging the pores of the earth and preventing the recharging of the water table. As a result, river water slips over the surface and fails to enter the bowels of the earth, causing further depletion of ground water. Third, the Kali Bein is a holy river, being the first pilgrimage for the Sikhs who visit Sultanpur Lodhi from distant places across the world to take a holy dip and drink the water of the river, believing it to be sacred and pure. They also use the same water in religious rites and for preparing food for religious gatherings. Pollution of the river's water is thus hurting the pilgrims, in terms of both their faith and their health (Nirmal Kuteya 2004).

The Indian constitution makes provision for the prevention of river pollution according to a law in effect since 1975. The Punjab state government also assured support to Baba Balbir Singh Seechewal's efforts, with statements to this effect by the chief ministers and prohibitory orders repeatedly passed to ban the flowing of sewage water into the river. In spite of these legal provisions and initiatives, the river's condition continued to deteriorate, and on May 5, 2006, Baba inspected the river with local city officials and identified the places at which sewage water was flowing into the sacred river. It was found that sewage water from dozens of villages in Kapurthala and Hoshiarpur districts was still flowing into the Kali Bein. This observation was summarized as a project report and submitted to the chief minister of Punjab on May 30, 2006 (Nirmal Kuteya 2004).

Another problem is that of encroachment around the Kali Bein area. The first type of encroachment comes from powerful construction industry leaders. Due to the rapid growth of population in the Doaba region and increasing scarcity of the land, the empty area on the banks of the Bein is grabbed for constructing unauthorized buildings by vested interests. The other type of encroachment is by farmers already owning lands adjoining the Bein who keep expanding their existing farms by grabbing the empty land around the Bein. They even influenced government officials to update the land records with their newly grabbed land pieces. This process went unchecked because, in most cases, revenue records of dimensions of the Bein area were not available. But where revenue records were available, some revenue authorities registered illegal sale deeds of encroached lands in favor of encroachers. Those who want the law to be implemented and the Bein areas to be vacated are left helpless due to powerful vested interests. Due to such encroachments, the passage of the Kali Bein was narrowed, leaving no space for its water to flow, resulting in frequent floods in the area during the monsoon (Nirmal Kuteya 2004).

While pollution and encroachment are common problems to all Indian rivers, the problem of scarcity of water is peculiar to the Kali Bein. There are three reasons for this. First, as mentioned above, the Beas, which used to be the main source of the Kali Bein, has moved away from the place of origin of the Bein over time. After parting with it, for some time, it went on supplying its water through underground flow. But with the establishment of Pong Dam in 1960, the level of water in the Beas itself went down, and it could no longer supply sufficient quantities of water for maintaining regular flow in the Bein. Second, the water level in Terkiana Marshland, which also fed the Bein, went down. Third, due to the wrong policy of local authorities, the gates of Budho-Barkat Barrage, which was erected in 1932, were closed and the store of water deposited behind the barrage was not allowed to flow naturally into the Bein but was directed towards the Beas again, depriving the Bein of its natural right of this water. Moreover, the passage of the Bein was filled up with silt flowing into it from rainwater brought by small seasonal brooks from the Bein's catchment area in Hoshiarpur district. The thick layer of silt chocked the pores of the earth. The Bein could not give outlet to huge stores of underground water, which forced its way out wherever it could, for example, in the fields submerging the crops growing there. Thus, agriculture in this area was the worst sufferer on this account. Lack of sufficient water in the Kali Bein worsened its already miserable condition. Due to its shortage of water, the flow of the Bein lost its force. At some times of the year, it came to a standstill and, mixed with polluted waters, started emitting a foul smell that polluted the surrounding environment. This led to the growth of the weed in the form of hyacinth. At the places where it dried up altogether, the speed of encroachment was further accelerated. It was turned into a dustbin for dumping the dirt and wastage of nearby towns and villages (Nirmal Kuteya 2004).

Restoring the Kali Bein

In 1969, on the fifth centenary of Guru Nanak, Sant Kartar Singh attempted to construct facilities for holy baths on the banks of Kali Bein. In 1983, ex-Maharaja of Kapurthala Sukhjit Singh and the *Manav Vikas Manch* (Human Development Forum, an NGO in Kapurthala) tried to raise the awareness about cleaning the Kali Bein. In 2000, V. K. Singh, the then administrator of Kapurthala, in association with the Guru Nanak Sacred Bein Restoration Committee, tried to get some water released into the Kali Bein from the Mukerian Hydel Channel. NGOs such as Dharat Suhavi Chowgirda Bachao Committee and Punjabi Sath Lambra also tried to organize meetings to protect the Kali Bein. However, all these efforts remained intellectual exercises with very limited actual transformation of the Kali Bein (Souvenir 2006).

Baba Balbir Singh Seechewal, who had already earned a name of *Sarakon Wale Baba* (Baba of the roads) due to his great work to construct roads across the state of Punjab, was invited to a meeting on July 15, 2000, to discuss the plight of the Kali Bein. He quickly realized the futility of mere words, and the very next day arrived at Sultanpur Lodhi and plunged into the action of cleaning the river himself. Soon, it turned into a mass movement with thousands of people volunteering to join the effort.

From its beginning in July 2000 to April 2003, the *Kar Sewa* (volunteer work offered as devotion) focused on cleaning the holy rivulet at Sultanpur Lodhi. At that time, the Kali Bein's plight was at its worst with solid waste, dead bodies of cattle, and sewage water of the town being thrown into it directly. People also abused the river for their daily secretions. Without any proper road on its banks, the area surrounding the river used to be deserted for most of the time. One of the earliest efforts involved building roads, streetlights, and platforms around the river. From March 30, 2003, to April 1, 2003, a campaign was organized to convince the people not to drain their wastewater into the Kali Bein. The pioneering villages, which promised not to pollute the river, were publicly honored. From May 2003 to May 2004, massive cleansing efforts by thousands of men, women, and children, led by Baba, succeeded in cleaning the river from Dhanoa in Hoshiarpur district to the Kanjli wetland. During this project, the bed of the Bein was cleaned and waterweeds were removed. The river was full of snakes and many volunteers were bitten by them. Fortunately, the work continued despite these problems and dangers. A paved pathway was constructed from Dhanoa to Kanjli measuring 110 kilometers. Every July, the anniversary of Kar Sewa is celebrated with several religious ceremonies, sports tournaments, and seminars with scholars, officials, politicians, journalists, and religious leaders. Such celebrations are used to spread further awareness among the people about the environmental problems and their solutions. Since 2003, large religious singing processions (*nagar kirtans*), especially on the birth anniversary of Guru Nanak, are organized to spread environmental awareness. In January 2004, Baba launched another series of construction projects at the bathing platforms at Gallowal. In April 2004, cleaning and construction work continued at Bhulath, Subhanpur, and Kanjli. On April 14, 2004, Sikhs celebrated Vaisakhi, a major Sikh festival, on the banks of the Kali Bein. Thousands of people took the holy dip in its water at the newly built bathing platforms. Baba addressed the gathering and emphasized the religious and environmental significance of the river and the land. Since 2004, Vaisakhi has become another major annual holy celebration in which people are reminded of their environmental responsibilities (Souvenir 2006).

From January 2005 to May 2005, similar bathing platforms and flowerbeds were built at Sultanpur Lodhi. Several plants were also sown to the west of Talwandi Bridge. In June 2005, the river was further beautified with shelters for the pilgrims. The Talwandi Bridge near Gurudwara Ber Sahib was renovated as well. The lighting arrangements were also installed here on May 2006. In December 2006, the Kar Sewa entered a new phase, with the focus shifting to downward Gurudwara Ber Sahib for cleaning and deepening the riverbed until Harike Pattan, about 35 kilometers away from Sultanpur Lodhi, where the Kali Bein enters the Beas again. A large number of people removed the thick layer of silt, which opened the pores of the bed and promoted recharging of the water table, leading to deepening of the bed of the Bein. Water from the Beas came upstream to reach Sultanpur Lodhi on June 24, 2006. On January 8, 2006, three boats were inaugurated for the entertainment of visitors to the Kali Bein, which were donated by Harjinder Singh of the UK. In May 2006, the river was further cleaned at Mukerian Hydel Channel. Due to lack of water in the river, hyacinth, which grows quickly, had to be cleaned repeatedly at various places. In June 2006, the cleaning work began at Himmatpura in Hoshiarpur district (Souvenir 2006).

In addition to working on the ground, Baba Seechewal also submitted several petitions and appeals to government officials. On January 3, 2005, he gave a memorandum to deputy commissioner of Kapurthala appealing him to protect and preserve the Kali Bein by taking the necessary steps. On December 10, 2004, he held a meeting with several administrators on a similar theme. On November 28, 2005, a large number of people led by Baba Seechewal met other administrators for the same cause. Since 2005 onwards, Baba held several press conferences to reach out to the masses with similar appeals. The four demands Baba has raised with the Punjab government are as follows (Souvenir 2006):

First, the area of Kali Bein should be demarcated as per revenue records, the illegal possessions be vacated, and all illegal sale deeds in the area should be canceled. Second, a green belt should be declared up to 250 meters on both sides of the Bein, where construction of houses and factories would be prohibited. Third, at least 400 cusecs of fresh water should be released into the river from Mukerian Hydel Channel, so that its flow becomes regular. Fourth, considering its historical and religious significance, Kali Bein should be declared a holy river and Sultanpur Lodhi should be declared a holy town. The Punjab State Government declared them both holy in 2006. However, other demands have yet to be implemented on the ground, although they are accepted in principle by the government (Souvenir 2006).

Dharma and science of Baba Seechewal's work

According to Baba, with the advent of modernity, people stopped the use of traditional means of water such as wells, ponds, and rivers and adopted the modern means of water such as tube wells, motor pumps, and submersible pumps. In older times, people respected the water sources and even revered them, realizing their significance for life. Fairs and festivals were celebrated on the banks of rivers and other water sources. Adulations were offered, rituals were performed, and free food used to be distributed on the banks of these rivers. Sikh holy text Guru Granth Sahib gives water the sacred status of a deity at many places,[1]

> The corn is sacred, the water is sacred, and the fire and salt are sacred as well. When the fifth thing, the ghee is added, the food becomes pure and sanctified.
>
> (Section 08 – Raag Aasaa – Part 127, p. 473)

> Air is the guru, water is the father, and earth is the great mother of all. The air that we breathe in, that is the basis of our life and through which the Word is uttered, is called the Guru in the Gurbani. Water is called the father and the earth the mother.
>
> (p. 8)

> First, there is life in the water, by which everything else is made green.
>
> (Section 08 – Raag Aasaa – Part 126, p. 472)

> In the bowl of the sky, the sun, and the moon are the lamps; the stars in the constellations are the pearls. The fragrance of sandalwood is the incense, the wind is the fan, and all the vegetation is flower in offering to you. O Luminous Lord! What a beautiful lamp-lit worship service this is! O destroyer of fear, this hymn is in your service.
>
> (p. 663)

In addition to quoting from Sikh holy texts, Baba also cited the examples from the lives of the Sikh gurus. The sixth guru Hargobind Sahib established the holy town of Kiratpur Sahib with several parks and gardens. This guru also advised Mughal emperor Jehangir about planting trees by the roadside as an important duty of an ideal ruler (Macauliffe 1963). Baba cited the example of Guru Nanak who broke the pride of Wali Kandhari, who asserted that water was his private property, but Nanak taught that it is a divine gift to be shared with all living beings. Similarly, Baba cited the example of the seventh Sikh Guru Har Rai Sahib, who had inspired his followers to plant trees.

Baba also explained the process of water purification in a scientific manner in this way. First, store the sewage water of each town at a proper place depending on the availability of space and other facilities. By using sedimentation and decantation processes, it can be made usable for agricultural purposes. The dirty silt that settles down at the bottom can prove a very useful fertilizer. A pipeline is required to carry this dirty water to the agricultural farms. Unfortunately, in our present scheme of water management, the dirty water is not reaching the place where it is required but it is reaching elsewhere. Almost all the villages and towns are situated on high grounds. Ironically, first the sewage water is gathered behind dikes and dams to flood the town and then it is drained to pollute the same water. A number of small treatment plants should be set up at different places in every town and treated water should be supplied to the farmers for irrigation. Unfortunately, the current system drains out the dirty water into the rivers instead of treating and then supplying it to the farmers. The dirty water thus entering the rivers is not only polluting the water for the present generation but is also entering the ground water and polluting the water for future generations. Moreover, inflow of effluents in the rivers creates a thick layer of silt that settles down at the bottom. Because of this hard crust under the bed, the pores of the soil are blocked and the water of rivers, lakes, and rains cannot recharge. The silt that has settled down under the bed of rivers should be cleaned out so that the pores of the soil are opened for the water to descend deep into the earth. Temporary dikes should be erected so that rainwater may be blocked for free seepage. This will help raise the water table. Speaking from his own experience of installing water treatment plants at some places, Baba further explained that they have tested the efficacy of this method at many places with the cooperation of the people and found that this dirty water was beneficial for crops (Nirmal Kuteya 2004).

Unlike some traditional religious leaders, Baba does not view science as an enemy to religion. According to him, "both religion and science are complementary to each other and science needs to be religious and religion needs to be scientific. Sociocentric science is religious and rationally practiced religion devoid of superstitions is scientific." At every step of his social work, Baba used various scientific tools and techniques, including indigenous and foreign technologies. According to Baba, any technology that can be used for social service is acceptable. Also, that technology should be practically feasible for the social, economic, and ecological conditions in Punjab. Every imported technique must be fine-tuned for the local conditions before its application. In different types of Kar Sewa, Baba and his colleagues and followers used different types of tools such as sickles, blades, spades, land-driven levelers, tractors, and cranes. Using machines and other techniques, Baba and his team learned their shortcomings and kept improving them. In the process of their new experiments, they kept developing newer techniques, methods, and implements that proved more efficient, affordable, and suitable for local requirements and resources. For instance, a new blade-like instrument was devised to clear the weeds from the bed of the Kali Bein. It proved extremely useful in cutting out the weeds from their entanglements. A lot of spadework was done on the Kali Bein and the roads. Baba and his team devised a new kind of more efficient and convenient spades to unload the trolleys full of soil, level the ground, dig ditches for sowing plants, and lay foundations for buildings and bathing platforms. Hand-driven levelers (*Jandra*) were devised and used to drive the unnecessary soil off the road, as well as to level the ground for making the bricked road. A tractor-driven soil reaper (*Jholi Wala Karah*) was used on the Kali Bein and on the roads for transferring the extra soil. A special raking implement was also devised for extracting hyacinth plants out of the Kali Bein. Under the bed of the river, a hard crust of soil had formed because it had remained undisturbed for a long time. The link between the surface water and the underground was broken. In the waterlogged areas, it was necessary to revive the link for the underground water to come out and flow into the Kali Bein. In dry areas, this link was necessary for facilitating the recharging of underground water. In order to revive this link, heavy tractors were used inside the river to upturn the hard crust. Some of the latest cranes, donated by a UK-based Sikh donor, were used on the Kali Bein for desilting the bed of the Kali Bein, preparing paths, digging for installing the sewerage pipes, and loading the soil into the trolleys. Similarly, boats, also donated by UK-based Sikh donors, were used both for cleaning work and for recreational purposes. At Seechewal, a permanent workshop is maintained where a team continues to experiment with newer tools and technologies. One of Baba's disciples, Sukhjit Singh, renounced his government job as a mechanical foreman and became a full-time associate of Baba. He now supervises the technical experimentation at Nirmal Kuteya, headquarter of Baba's activities. All such experiments have led to putting lintels without steel over the roofs of the buildings that are stronger and less expensive. Similarly, in construction of the bathing platforms along the Kali Bein, techniques and designs of laying the bricks and stones are decided by Baba himself. Also, Baba and his team laid out an efficient and affordable sewerage system and waste treatment plant for Seechewal and other nearby villages.

I also noticed widespread use of latest media at various places, including an FM radio station, filmmaking center, and multimedia computer laboratory (Nirmal Kuteya 2004).

Impact of the Kar Sewa

As a result of the Kar Sewa, the Kali Bein has been cleaned and the flow of water has been restored, and the waterlogged areas of Hoshiarpur have an outlet for excess water. Agriculture has been made possible in large areas of Dasuya and Mukerian, leading to widespread prosperity for the farmers. Due to the stoppage of the flow of water in the Kali Bein, a vast area in Kapurthala was affected by the depleting water table. After the restoration of flow, the recharging of water has started and the water table is rising again. As a result, there are signs that the imminent danger of the desertification of land is averted and the wells have been recharged again. The quantity of water being released into the Kali Bein from the Mukerian Hydel Channel is still very small, not enough to reach Sultanpur Lodhi. To solve this problem, Kar Sewa brought the water from the Beas upward from Harike Pattan, resulting in better recharging in Sultanpur area (Souvenir 2006).

The Kar Sewa has resulted in massive environmental awareness in Punjab's Sikh community and the Sikh diaspora across the world, and many villages and towns have stopped polluting the river since then. The newly built road from Dhanoa to Kanjli has facilitated transportation in the area. Now the farmers can reach their lands easily and can take their produce easily to the markets. This has led to steep growth in the land prices in the area and a further boost to the local economy. The Kar Sewa has revived the religious significance of the Kali Bein as the birthplace of the Sikh philosophy and the opening verse of the Sikh holy text. The newly built bathing platforms at five places have helped people perform their rituals on the banks of the river in greater number and greater convenience. Efforts to keep the river clean and stop the inflow of dirty water have started the process of development, with many villages installing sewerage systems and waste treatment plants (Souvenir 2006).

The river at Sultanpur Lodhi is now clean, with hyacinth and silt removed, dikes made, platforms built on its banks, trees and plants planted on its sides, permanent water and lighting arrangements made, and seating and boating arrangements made for pilgrims and tourists. Baba also conducted large programs for planting trees on the roadsides, riverbanks, courtyards, cremation grounds, and wastelands. In many villages, dirty ponds were transformed into beautiful parks. In his discourses and lectures, Baba preaches about tree plantation and distributes tree saplings to devotees as *Prasad* (graceful gift from the divine), with exhortations to take care of them until they grow as trees. He also organizes tree-plantation camps in various villages and towns in association with local administrators (Souvenir 2006).

Recognitions and honors for Baba's work

One of the most prominent visitors to see Baba's efforts were India's President Dr. APJ Abdul Kalam, who had these words of praise in one of his speeches given on the occasion of National Technology Day on May 11,

2004. He later visited Seechewal and surrounding areas on August 17, 2006, and July 27, 2008, to see Baba's work (*Souvenir* 2006).

Baba organized people's participation in stopping the massive flow of sewage into the Kali Bein and cleaned 160 kilometers long polluted and chocked rivulet by deploying on average 3,000 volunteers workers per day. Today one can feel the flow of fresh water in the rivulet released from the Terkiana Barrage by the government. The revival of the rivulet has recharged the water table as the hand pumps that had become dry for the past four decades are now pumping out water. Baba not only did the cleaning up operation by clearing Bein from the weeds and hyacinth, but also built bathing platforms at many places. He also built more than hundred kilometers long road on the bank of rivulet. While I was thinking how we should solve the problem of improving the environment of rivers and religious places, I find one of our enlightened citizens has taken the initiative and demonstrated the power of ignited individuals to solve a societal problem. Let this model spread in all the places of divine worship and inspire the pilgrims to participate in the divine task of clean environment in water and air. Thousands of local initiatives will lift India up and make it green.

In addition to such accolades received from the Indian President, Baba's socio-environmental work has also been reported in prominent Indian media outlets such as *Times of India, Hindustan Times, Outlook India, NDTV, IBN TV, Tribune India, Deccan Herald, The Hindu, Rediff,* and *Indian Express.* *Time Magazine* chose Baba as one of the Heroes of the Environment in 2008 (Nirmal Kuteya 2004).[2]

Note

1 https://www.sacred-texts.com/skh/granth/gr08.htm (accessed June 7, 2014)
2 http://content.time.com/time/specials/packages/article/0,28804,1841778_1841781_1841808,00.html

6 Dharma of GEP and science of GDP

Gross domestic product is considered the most widely used standard to measure the development rate of any country. A country's growth is judged by its GDP, which primarily is calculated based upon the growth of its industries, real estate, and other service-related industries. All of these sectors are considered "produce" for the measure of a country's GDP. However, literally, only minerals and agricultural produce should be considered the "produce" of a country. Only the products that are derived from agriculture and from mines are required for human survival and its basic needs, whereas all the other so-called products tend to be used for our luxury and convenience. Ironically, in most of the discussions and discourses of GDP, a country's environmental and agricultural resources are barely mentioned. Perhaps, because of this skewed emphasis on industrial and economic growth, environmental growth is neglected.

A developing country such as India, where 85% of its people can afford only bare necessities, a growing rate of GDP has little significance. GDP, as Joshi argues, is instead a comprehensive index of unsustainable growth largely based on industrial growth and infrastructural development. The beneficiaries of this latter kind of growth make up the top 15% of India's population. Moreover, the current GDP sectors are harming the rest of India's population. Increasing impetus towards measuring the GDP creates the illusion that GDP is the only measure of a country's development and ignores the fact that it lacks the comprehensive, inclusive, and sustainable aspects of true development. And in the name of growth, basic raw materials such as minerals and forests and natural resources such as air and water continue to be overshadowed and ignored. The economy is equated with the finished products coming out of industries, which are highlighted in the stock markets and which in turn are treated as the barometer of a country's economy.

Especially in last two decades, the growing GDP in India (and elsewhere in the world) has adversely impacted its environmental resources. India's air, water, and forests continued to worsen, resulting in direct consequences of climate change, such as warmer summers, late monsoons, colder winters, intense floods, drying and polluted rivers, loss of fertile lands, unclean urban air, and decreasing forest covers. Corporations saw a business opportunity even in this crisis, and bottled water and soft drinks continue to be a growing business in India, putting further pressure on fresh water resources. Similarly, in the name of increasing land productivity, chemical fertilizers and pesticides have wreaked havoc in Indian states such as Punjab and elsewhere, where topsoil is destroyed and even natural forests show signs of this chemical invasion. Similarly, rising demands for furniture, automobiles, electronics, and other appliances from a growing middle class continue to threaten the remaining forests, mines, and mountains of India.

The massive exploitation of natural resources has been acknowledged at United Nations conventions on climate change, such as in Kyoto, Copenhagen, Cancun, and Doha. On one hand, the increasing focus on and the increasing rate of GPD by the developing countries, such as Brazil, Russia, India, and China ("BRIC"), have emerged as a cause of concern for developed countries from the climatic perspective because of its resultant rise in CO_2 emissions. On the other hand, developing countries, led by BRIC, have continue to resist such pressures from the developed countries and have argued oft-repeated phrase "development before environment," for example, as mentioned by India's Prime Minister Indira Gandhi in one of the earliest UN Environmental Conference in Stockholm in 1972. In the name of environment, the real battle that continues to be waged by both sides is about the economic and industrial growth while the real raw materials and bare necessities of life such as air, water, land, forests, and mines, continue to be neglected by policymakers on both sides. In addition to GDP as the measure of growth, there can be a comprehensive measure of ecological growth, and then perhaps this current climatic crisis can be understood as the first step towards its resolution.

The current climatic crisis is the result of mindless exploitation of natural resources. As an example, India's rivers – both seasonal and glacial – have consistently lost their water in terms of both quality and quantity. The biggest reason for the decline in water level in these rivers is the loss of forests in their territories, which are an easy target for the rise in consumption fueling the industrial growth as measured by GDP. Similarly, in last few decades, topsoil has become heavily polluted and mixed with chemicals, which has not only impacted human health but also damaged environmental health. Emissions from industries, automobiles, and air conditioners have polluted the air in the same ways. In metropolitan cities like Delhi and Mumbai, electronic billboards continuously display the levels of SO_2 and CO_2 polluting the air. It is indeed unfortunate that a development that destroys its very foundation, its raw materials for life and for economy, has emerged as the singular goal of most national policymakers in India (and elsewhere). The nature-gifted resources such as air, water, forests, and mines have been unfortunately taken for granted, and the so-called GDP-based development is benefitting only 15% of India's population. This "India Shining" scenario was categorically rejected by Indians, for instance, in the 2004 national elections when the ruling BJP party could not get enough votes for its second term at the central government.

Forty-five percent of Indians even today lack the bare necessities of life, such as permanent employment and a house with electricity and sanitation. In such a scenario, India's development cannot be reflected based solely on the rise in its GDP as often portrayed in national media or government proclamations. Common Indians' survival directly depend on basic food, housing, clothing, air, water, electricity, and employment. The majority of Indians in its thousands of villages live quite like subsistence-based indigenous communities of Native Amerindians. Nature is their only resource for their daily requirements of food, fuel, fodder, and fertilizer. None of this matters, at least not in any significant way, for the measurement of GDP, and this neglect continues to negatively impact India's natural resources.

The scarcity of natural resources has not just affected rural livelihood, but also national and international ecology. Thus, it is imperative that the definition of development should include not just industrial and infrastructural development, but also the development of essential natural resources. This inclusion will go a long way in ensuring the balance between economic and ecological development. Instead of GDP (gross domestic product), we should design and deploy a new index – GEP (gross environmental product) to measure both the economic and ecological development of a country. This will help us determine the direction of our development and its sustainability. Otherwise, the chasm between GDP and GEP will continue to widen, eventually destroying the very foundations of GDP. It is high time to replace GDP with GEP in all national and international forums and conferences.

According to a news report (Mahapatra 2015),[1] HESCO took a concrete step to pursue this line of thinking and filed the first ever public interest litigation (PIL) in Uttarakhand High court in 2011 on making GEP "a co-indicator of the state of the economy along with the standard GDP". Agreeing with this petition, the stage government agreed to incorporate environmental factors while measuring GDP.

Even as HESCO challenged the state government, India's national minister for environment and forests, Jairam Ramesh, was establishing an expert group to transition India from GDP to a Green GDP model.[2] Ramesh, addressing the India Today Conclave's tenth anniversary, acknowledged that the environment entered this annual meeting of influential Indians only in 2011. Citing reasons of livelihood, public health, climate change, and sustainability, Ramesh unveiled his plans for India to move away from GDP to a Green GDP. He announced that Professor Partha Dasgupta, an eminent environmental economist at Cambridge University, had accepted the offer to lead an expert group who will ensure this transition. In April 2013, the expert group presented its progress that it is working to establish National Accounts to measure the stock of natural, human, and physical assets of the country. As Dr. Joshi hopes,

> The world will not transform with HESCO's 1,000 watermills or 200 food-processing units. The biggest challenge is economic disparity and ecological imbalance. They both are two sides of the same coin. HESCO's work involves both. After more than thirty years of work, we have concluded that the rural community will survive only with Gross Environmental Product, not Gross Domestic Product.

Notes

1 http://www.downtoearth.org.in/content/true-measure-growth
2 "The Way to a Green GDP" delivered at India Today Conclave, New Delhi, March 18, 2011. http://moef.nic.in/downloads/public-information/Way%20to%20 Green%20GDP.pdf

7 Conclusion

Is there a discrepancy in the way that many non-Western peoples, especially of the global South, view the scientific and religious worlds and the way these spheres are viewed in the Westernized world? This book has attempted to present a few examples, such as those of rural people in the Indian Himalayas, who see the scientific and religious words as overlapping and intertwined. Like other non-urbanized, non-industrialized, and non-Westernized people, Himalayan villagers consider animals and plants their relatives, and hold several rituals and beliefs that actively incorporate all natural beings and resources. The worldviews and categories of the natural world by non-Western peoples of the global South still are not included in Western contexts, which continues to be ontologically dichotomized in scientific versus religious worlds (Veteto 2013).

According to anthropologist Julie Cruikshank, as cited by Veteto (2013), indigenous people in Northwest Canada and Alaska believe that disasters can occur if a glacier is insulted. They also have a taboo on cooking with animal grease as a mark of respect for glaciers. For the modern Western knowledge categories, such beliefs remain a mystery. Even celebrated environmentalists such as John Muir have ignored such traditional knowledge because, according to Western ontological and epistemological categories, glaciers and other natural entities cannot have sentience and agency. However, it is becoming increasingly obvious that modern Westernized and urbanized societies have an unsustainable relationship with the natural world, resulting in the threat to the very survival of our planet. In contrast, smaller rural and indigenous communities in their respective parts of the world have been maintaining a sustainable relationship with nature for thousands of years.

As Veteto shows (2013), in recent years, anthropologists and other scholars have worked with various rural and indigenous communities and have shown that Western scientific knowledge is inadequate in its categories of "natural" and "cultural" and cannot do justice to indigenous relations, such as with glaciers in North America. Researchers have similarly discovered numerous agroforestry tracts and mixed-use orchards that were wrongly classified by scientists as old-growth forests. For instance, anthropogenic dark earth soils continue to be found deep inside the Amazonian rainforest, proving that the majority of American landscapes were managed sustainably by indigenous communities for millennia. Contrary to the European explorers' assumption that the "new world" was a sparsely inhabited wilderness, the Western hemisphere has always been, as proved by historical ecologists, an intensely managed landscape for millions of indigenous people. The Aztec capital city of Tenochtitlan and several Asian cities were more developed than any European cities when they were first "discovered" by the colonial powers. The Asian and American landscapes and local ecologies were being sustainably managed by indigenous communities using traditional ecological and scientific knowledge systems, and European ("Western") scientists could not, and still cannot, classify Asian and American phenomena in the strictly dichotomized cultural and natural worlds, two categories that are still juxtaposed on the non-Western peoples. It is this traditional ecological knowledge that this book also focuses on, and the work of HESCO makes us reconsider our categories of science versus religion and nature versus culture.

Non-Western traditions and communities, such as those of India, are much better positioned to provide an alternative to Western dualistic thinking. Scholars generally trace the nature–culture and science–religion divide back to Descartes and then provide examples to challenge this dichotomy. However, as Veteto (2013) argues, Descartes does not bring a new idea due to the modern "industrial revolution" in Europe, but he is merely a culmination of an ontological divide that started as early as in the Aristotelian thought. Therefore, this dichotomy that is at least 2,000 years old is now being enforced on the rest of the world due to colonization and globalization.

Sandra Harding (2009:19) notes that the postcolonial movements, like HESCO,

"expand the Copernican revolution as they decenter and parochialize dominant ways of thinking about the production of scientific and technological knowledge and their familiar philosophic assumptions"[1] some of which I reviewed above. All the HESCO projects and experiments described above do not advocate that "non-Western cultures in effect 'delink' from Western sciences and develop regional sciences out of the fertile ground of their own traditions"[2] (Harding 2009). Instead, HESCO "clearly recognizes that, on the one hand, one cannot simply abandon modern Western sciences and their philosophies. On the other hand, the Western technologies can be radically transformed through integration with regional legacies so as to enable the flourishing of a multiplicity of knowledge-traditions and the societies that depend upon them"[3] (Harding 2009:17).

This hybridity of Indian modernity is also noted by Joseph Alter (2004) citing Gyan Prakash (1999:234):[4]

> Indian modernity has always existed as an internally divided process. An aura of dislocation and disorientation has always accompanied Indian modernity's existence. I do not mean this in a negative sense; uncertainties and estrangements point to its source of creativity and to possibilities of new arrangements and new accommodations. There is simply no way to tidy up this messy history of India and narrate it as the victory of capital over community, modernity over tradition, West over non-West. These neat oppositions exist side by side with the history of their untidy complicities and intermixtures.

As Raman in his article (2012) notes,

> Science and religion are in harmony and in "conflict in the Hindu world, as in all dynamic civilizations. Hindu worldviews are rich in insights and flexible in their capacity to incorporate new ideas and grow. Even religiously inclined thinkers can accept evolution and other scientific paradigms, and even though astrology and numerology continue to impress and guide the masses, by and large they do not influence decision makers regarding the place of modern science in colleges and universities. In spite of attempts by some groups, educational institutions have thus far been shielded from such ancient sciences creeping into the curriculum. Perhaps the greatest strength of the Hindu framework lies in the fact that those who are brought up in the tradition generally have the capacity to accommodate contrarian worldviews that have contextual appeal and significance. The cheerful practice of traditional modes in the religious context gives meaning and purpose to most Hindus; and the willing embrace of rational science gives them a fuller appreciation of the natural world. Given the rich intellectual heritage, it is not surprising that when the Hindu world encountered what was then largely Western science, it received and absorbed it with eagerness. There are at least two reasons why very soon Indians of Hindu heritage began to accept the scientific framework and to participate productively in the international endeavor we call modern science. First, the Hindu psyche has always been accustomed to new and divergent perspectives, and it accepts multiplicity as intrinsic to the human condition. The second and no less important reason is that there has never been an institutionalized supreme authority, elected or anointed in the Hindu world, to dictate to its adherents what should and should not be accepted, what can and cannot be regarded as the Truth. This implies an extraordinary freedom in the religious context. Therefore, while rigid orthodoxy and what one would call fundamentalist doctrinal believers are still common in the

Hindu world, there are also open-minded religionists, free thinkers and atheists in the tradition, as also enlightened bridge-builders between time-honored culturally meaningful worldviews and impersonal scientifically derived results.

Some glimpses of such openness I noticed in the HESCO's work in Indian villages. Although HESCO's founders and other volunteers are practicing Hindus in their private lives, they are also thoroughly trained scientists and have thus evolved their methodologies by bridging the two worlds successfully.

HESCO's work with Himalayan communities is the Indian counterpart of what can be called social work. When I asked HESCO about the practical challenges or difficulties related to such work, they noted several challenges, one of which is how to sustain the transformation based on the HESCO's projects. Without the dharmic perspective, the work can become "mechanical." If these activities fail to inspire people to develop an ethos or bonding among the village communities, or if they stop practicing the ethics in their daily lives, then the work can take a "religious" shape based on the devotional faith of the words of Joshi. That would reduce the dharmic work into another religious ritual, as has happened with some movements in both India and elsewhere. Another challenge is to take these activities and replicate them at a higher level. So far, these have remained smaller local role models at the district level rather than projects at the regional or national level. They also confessed that the number of volunteers available to work at different sites varies according to the intensity and depth of HESCO's ideas in the surrounding villages. Since the spread of HESCO is not uniform across the different villages and towns of Uttarakhand (and elsewhere in India), the number of coordinators working at such projects is also not uniform. Noted social activist Anupam Mishra aptly remarked (1993):

> [P]eople are bringing out miracles at the grass-root level. Upon seeing them, we should humbly accept them. Even if they may not fit our measuring scale, our measuring scale itself may be inappropriate. A work that has already reached millions belongs to the people. Media reports only political parties but it cannot represent the people.

HESCO has developed several more projects to fructify its mission for social transformation, but I have described only some of them. These projects do not label themselves as "social projects" and yet they have succeeded in bringing the upper and lower castes together. Like any other such work, the challenge now is to maintain them and to develop new such projects. To my knowledge, the current leadership is focusedon strengthening the existing projects by inspiring more people to join their movement.

Finally, I would like to point out that perhaps the best way to individually break down the religion–science divide in Western society is by way of direct experience or learning from individuals from other cultures such as the Indic communities of HESCO and others who do not recognize this ontological dualism. I'm not sure that I could truly understand the issue had I not experienced particular ceremonial moments in such Indic communities, engaged in deep meditations with Hindu and Jain practitioners. Not everything can be accomplished rationally, and as Lao-tzu famously said, "The Way that can be named is not the Way." In many cases, I happened to be the right individual at the right place at the right time to have some of the experiences that I have; others were more widely accessible, and still others were a result of long-term trusting relationships built in a particular cultural community. In all cases, sacred ceremonies of other cultures should never be appropriated for individual or economic gain. Another good way of challenging engrained Western ontologies is by learning from scholars who come from non-Western traditions.

Although each has its own historical trajectory and culturally specific beliefs, Native American and Daoist; strains of Hinduism, Jainism, and Buddhism; and many other indigenous traditions all have unique approaches that stress non-duality between subject and object, mind and body, nature and culture. It is my opinion that this approach toward reality, each manifesting in its own distinct way, is a more accurate way of understanding our particular universe and reality. Many scholars have made the case that insights from these and other cultural traditions are more consistent with the deepest insights of our own theoretical physics – but in our everyday lives and institutions, we remain largely trapped in a Newtonian universe of physical laws guided by the ontology of Plato and Descartes.

We need to get back to this pre-Socratic perception of nature–culture if we have any chance of solving our contemporary environmental crisis. This doesn't mean traveling back in time or trying to reconstruct the ceremonies of the Tuatha Dé Danann from thousands of years ago. If we can intellectually work through the nature–culture and science–religion questions and employ our insights in applied work, we may begin to help create a social reality that encourages the direct experience of non-duality. I see this as one of the most pressing goals for the Academy, and I think much can be accomplished by combining insights from non-Western communities and traditions.

Notes

1 Dipesh Chakrabarty, *Provincializing Europe: Postcolonial Thought and Historical Difference*, Princeton: Princeton University Press, 2000; Mignolo, *Local Histories*.
2 Samir Amin, Delinking: Towards a Polycentric World, New York: ZedBooks, 1990. Third World Network, 'Modern Science in Crisis'.
3 Dipesh Chakrabarty, *Provincializing Europe: Postcolonial Thought and Historical Difference*, Princeton: Princeton University Press, 2000; Mignolo, *Local Histories*.
4 Prakash, Gyan. *Another Reason: Science and the Imagination of Modern India*. New Delhi: Oxford University Press, 2000.

Bibliography

Agarwal, Anil. "You Are the Government", *Down to Earth*, Vol. 8, No. 5 (March 1999).

Agarwal, Rakesh. "Pickled Power", *Down to Earth*, Vol. 4, No. 2 (January 31, 1995), http://www.downtoearth.org.in/node/27430

———. "No Small Change" in Suhit Sen and Ruksan Bose (eds.), *Agenda Unlimited: Down to Earth*. New Delhi, India: Society for Environmental Communications, 2005, p. 233.

Agarwal, Sunil K. "Re-Energizing Watermills for Multipurpose Use and Improved Rural Livelihoods A Case Study from the Western Himalayas", *Mountain Research and Development*, Vol. 26, No. 2 (May 2006), pp. 104–108.

Alter, Joseph S. *Yoga in Modern India: The Body between Science and Philosophy*. Princeton, NJ: Princeton University Press, 2004, p. 76.

Anderson, Robert S. "Cultivating Science as Cultural Policy: A Contrast of Agricultural and Nuclear Science in India", *Pacific Affairs*, Vol. 56, No. 1 (Spring 1983), pp. 38–50.

Atkinson, Edwin T. *The Himalayan Gazetteer, Or, the Himalayan Districts of the North Western Province of India*. New Delhi, India: Low Price Publications, 2002.

Baker, Laurie. *Earth Quake*, LaurieBaker.net, 1992.

Balagangadhara, S. N. *The 'Heathen in His Blindness . . . ' Asia, the West, and the Dynamic of Religion*. Leiden, The Netherlands: Brill, 1994.

Balasubramanian, A. V. "Is There an Indian Way of Doing Science?" in A.V. Balasubramanian and T. D. Nirmala Devi (eds.), *Traditional Knowledge Systems of India and Sri Lanka: Papers Presented at the COMPAS Asian Regional Workshop on Traditional Knowledge Systems and their Current Relevance and Applications*. Chennai, India: Centre for Indian Knowledge Systems, 2006, pp. 183–192.

Berreman, Gerald D. *Hindus of the Himalayas: Ethnography and Change*. Berkeley, CA: University of California Press, 1972.

Bharati, Agehananda. "Actual and Ideal Himalayas: Hindu Views of the Mountains" in James F. Fisher (ed.), *Himalayan Anthropology: The Indo-Tibetan Interface*. The Hague, The Netherlands: Mouton, 1978, pp. 77–82.

Bhattacharya, Swapan Kumar. *Farmers Rituals and Modernization: A Sociological Study*. Columbia, MO: South Asia Books, 1976.

Bhawuk, Dharm P. S., Susan Mrazek, and Vijayan P. Munusamy. "From Social Engineering to Community Transformation: Amul, Grameen Bank, and Mondragon as

Exemplar Organizations," in Urbain, Olivier, and Deva Temple. *Ethical Trans-formations for a Sustainable Future*. New Brunswick: Transaction Publishers, pp. 75–112, 2011.

Bhowmick, P. K. *Applied-Action-Development Anthropology*. Kolkata, India: Institute of Social Research and Applied Anthropology, 1990.

Bilimoria, Purushottama. "Hindu Doubts about God", *International Philosophical Quarterly*, Vol. 30, No. 4 (1990), pp. 481–499.

———. "A Prolegomenon for Reconfiguring Science and Theology within Metaphysics, India and the Rest" in P. Bilimoria and M. K. Sridhar (eds.), *Traditions of Science Cross-Cultural Perspectives: Essays in Honour of B V Supparayyapa*. New Delhi, India: Munshiram Manoharlal, 2007, pp. 272–294.

———. "War & Peace between Science and Religion: The Divine Arch after the Four Horsemen", *Journal of Indian Council of Philosophical Research*, Vol. 28, No. 2 (April–June 2011), pp. 27–54.

———. "All India Radio: The War between Science and Religion in the Subcontinent – A Western Import?" in *American Academy of Religion, Chicago, November 2012* (Co-Sponsored Session: Religion in South Asia and Science, Technology, and Religion Unit).

Biswas, S. K. *Central Himalayan Panorama: Vol I*. Calcutta, India: Institute of Social Research and Applied Anthropology, 1993.

Brandtzaeg, Brita. "Women, Food and Technology: Case of India", *Economic and Political Weekly*, Vol. 14, No. 47 (1979), pp. 1921–1924.

Brown, C. Mackenzie. *Hindu Perspectives on Evolution: Darwin, Dharma, and Design*. London, England: Routledge, 2012.

Callicott, J. Baird. "A NeoPresocratic Manifesto", *Environmental Humanities*, Vol. 2 (2013), pp. 169–186.

Chakrabarty, Dipesh. *Provincializing Europe: Postcolonial Thought and Historical Difference*. Princeton, NJ: Princeton University Press, 2000.

Chopra, Ravi. *Survival Lessons: Himalayan Jal Sanskriti*. Dehradun, India: People's Science Institute, 2003.

Chowdhury, Salman. "Residential Project", *Doon School Weekly*, No. 2099 (October 15, 2005), p. 3.

"Community-Based Natural Resource Management in the Central Himalayas" in Ajit Menon et al. (eds.), *Community-Based Natural Resource Management: Issues and Cases from South Asia*. Los Angeles, CA: Sage, 2007.

Dharampal. *Indian Science and Technology in the Eighteenth Century: Some Contemporary European Accounts*. Delhi: Impex India, 1971.

———. *Essays on Tradition, Recovery and Freedom*. Goa, India: Other India, 2000.

Dhyani, R. P. *An Approach to Economic Planning for the Rural Poor of Central Himalayas*. New Delhi, India: Classical Publishing Company, 1994.

Dorman, Eric R. "Hinduism and Science: The State of the South Asian Science and Religion Discourse", *Zygon*, Vol. 46, No. 3 (2011), pp. 593–619.

Drew, Georgina. *Ganga Is 'Disappearing': Women, Development, and Contentious Practice on the Ganges River*. PhD dissertation, UNC Chapel Hill. 2011.

———. "A Retreating Goddess? Conflicting Perceptions of Ecological Change near the Gangotri-Gaumukh Glacier", *Journal of Study of Religion, Nature, and Culture*, Vol. 6, No. 3 (2012), pp. 344–362.

Duquette, Jonathan. "'Quantum Physics and Vedanta': A Perspective from Bernard D'espagnat's Scientific Realism", *Zygon*, Vol. 46, No. 3 (2011), pp. 620–638.

Eck, Diana L. *India: A Sacred Geography*. New York, India: Harmony Books, 2012.

Gantayet, L. M. *Chemical Science and Engineering*. Mumbai, India: BARC, 2007.

Geertz, Clifford. "Prophets Facing Backwards: Postmodern Critiques of Science and Hindu Nationalism in India", *Common Knowledge*, Vol. 13, No. 1 (Winter 2007), pp. 143–144.

Ghosh, Abhik. *Reinventing the Watermill in the Himalayas: The Gharat in History, Tradition and Modern Development*. New Delhi, India: Northern Book Centre, 2008.

Gibson-Graham, J. K. "Post Developmental Possibilities for Local and Regional-Development" in A. Pike, A. Rodriguez-Pose, and J. Tomaney (eds.), *Handbook of Local and Regional Development*. London: Routledge, 2010.

Gopalan, C. and Bani Tamber Aeri. "Strategies to Combat Under-Nutrition", *Economic and Political Weekly*, Vol. 36, No. 33 (August 18–24, 2001), pp. 3159–3169.

Gosling, David. *Science and the Indian Tradition: When Einstein Met Tagore*. London, England: Routledge, 2007.

Gupta, Akhil. *Postcolonial Developments: Agriculture in the Making of Modern India*. Durham, NC: Duke University Press, 1998.

Haberman, David. *People Trees: Worship of Trees in Northern India*. New York, NY: Oxford University Press, 2013.

Harding, Sandra. "Postcolonial and Feminist Philosophies of Science and Technology: Convergences and Dissonances", *Postcolonial Studies*, Vol. 12, No. 4 (2009), pp. 401–421.

Jain, Pankaj. *Dharma and Ecology of the Hindu Communities: Sustenance and Sustainability*. Aldershot, UK: Ashgate Publishing, 2011.

Jardhari, Vijay. *Pahāri Kheti: Kisānon kā Paramparāgat Vigyān*. Dehradun, India: Himalaya Trust, 2003.

Joshi, Anil P. *Degenerating Sustainability*. Dehradun, India: Samaya Sakshaya, 1999.

———. Sunil K. Agarwal, and Rakesh Kumar. *Mountain Technology Agenda: Status, Gaps, and Possibilities: Developing Appropriate Technologies for Sustainable Livelihoods in Mountain and Hilly Areas*. Dehradun, India: Bishen Singh Mahendra Pal Singh, 2006.

———. Rakesh Kumar, and Kiran Negi. *Offering: A Blessing in True Sense*. Dehradun, India: HESCO, 2011.

———. Kiran Rawat, and Bhawana Karki. "Millet as 'Religious Offering' for Nutritional, Ecological, and Economical Security", *Comprehensive Reviews in Food Science and Food Safety*, Vol. 7 (2008), pp. 369–372.

Joshi, P. C. "Technological Innovations in the Garhwal Himalayas: The HESCO Case Study", *Anthropologist*, Vol. 1, No. 1 (1999), pp. 81–85.

Kak, Subhash. "STS: Hindu Perspectives" in Carl Mitcham (ed.), *Encyclopedia of Science, Technology, and Ethics*. Detroit, MI: Macmillan Reference, 2005, pp. 915–920.

Kalidasa, Mallinatha and M. R. Kale. *Kālidāsa's Kumārasam?bhava, Cantos I–VIII, Complete*. New Delhi, India: Motilal Banarsidass, 1967.

Kapila, Shruti. "The Enchantment of Science in India," *Isis*, Vol. 101, No. 1 (March 2010), pp. 120–132.

Kashyapa, S. M. Ayachit, Nalini Sadhale, and Y. L. Nene. *Kashyapiyakrishisukti*. Secunderabad, India: Asian Agri-History Foundation, 2002.

Kochar, Rajesh. "Cultivation of Science in the 19th Century Bengal", *Indian Journal of Physics*, Vol. 82, No. 7 (2008), pp. 1003–1082.

Kumar, Anoop. "Going beyond Slogans", *The Doon City Chronicle* (1998), pp. 22–23.

Kumar, Rakesh and R. Dobhal. "Koti: A Developed Village" in Anil P. Joshi, Sunil K. Agarwal, and Rakesh Kumar (eds.), *Mountain Technology Agenda: Status, Gaps, and Possibilities: Developing Appropriate Technologies for Sustainable Livelihoods in Mountain and Hilly Areas*. Dehradun, India: Bishen Singh Mahendra Pal Singh, 2006.

Kumar, Saravana, Md. Arzoo Ansari, Archana Deodhar, and Vinod Singh Khatti. "Isotope Hydrological Study on a Few Drying Springs in Surla Valley, Sirmaur District, Himachal Pradesh", *Current Science*, Vol. 103, No. 1 (July 2012).

Lal, Vinay. "The Tragi-Comedy of the New Indian Enlightenment: An Essay on the Jingoism of Science and the Pathology of Rationality", *Social Epistemology*, Vol. 19, No. 1 (2005), pp. 1–14.

Macauliffe, Max. *The Sikh Religion: Its Gurus, Sacred Writings, and Authors*. New Delhi, India: S. Chand, 1963.

Mathur, Nayanika. "Effecting Development: Bureaucratic Knowledges, Cynicism and the Desire for Development in the Indian Himalaya" in S. Venkatesan and T. Yarrow (eds.), *Differentiating Development: Beyond an Anthropology of Critique*. London, England: Berghahn Press, 2012.

Mishra, Anupam. *Aaj Bhi Khare Hain Talab*. New Delhi, India: Gandhi Peace Foundation, 1993.

Nanda, Meera. *Prophets Facing Backward: Postmodernism, Science, and Hindu Nationalism*. New Brunswick, NJ: Rutgers University Press, 2003.

Nandy, Ashis. *Alternative Sciences: Creativity and Authenticity in Two Indian Scientists*. New Delhi, India: Oxford University Press, 1995.

Nasr, Seyyed Hossein. *The Need for a Sacred Science*. Albany, NY: State University of New York Press, 1993.

Negi, Vikram and R. Maikhuri. "Innovative Options for Sustainable Rural Development in Central Himalaya, India." Paper presented at The 2nd World Sustainability Forum. (1–30 November 2012), http://www.sciforum.net/presentation/904 (accessed 9 April 2013).

Nene, Y. L. "Environment and Spiritualism: Integral Parts of Ancient Indian Literature on Agriculture", *Asian Agri-History*, Vol. 16, No. 2 (2012), pp. 123–141.

Nicolaysen, Anna Marie. *Empowering Small Farmers in India through Organic Agriculture and Biodiversity Conservation*. Ph.D. University of Connecticut, 2012.

Nigah, Manpreet. *An Assessment of Seechewal Initiative in the State of Punjab, India: An Example of Community-Based Conservation?* Thesis-University of Manitoba, Spring 2008.

Nirmal Kuteya, Johalan. *Kar Sewa at the pious Kali Bein and other developmental work performed by Shri Maan Sant Balbir Singh ji and Gursangat*. Jalandhar, India: Ek Onkar Charitable Trust, July 2004.

Pant, B. R. *Women and Nutrition in the Himalayas*. New Delhi, India: Anmol Publications Private Limited, 2004.

Parāśara, Nalini Sadhale, H V. Balkundi, and Y. L. Nene. *Krishi-Parashara: Agriculture by Parashara: A Text on Ancient Indian Agriculture in Sanskrit.* Secunderabad, India: Asian Agri-History Foundation, 1999.

Prasad, Amit. "Beyond Modern vs. Alternative Science Debate: Analysis of Magnetic Resonance Imaging Research", *Economic and Political Weekly*, Vol. 41, No. 3 (January 21–27, 2006), pp. 219–227.

Rajan, S. Ravi. "Science, State and Violence: An Indian Critique Reconsidered", *Science as Culture*, Vol. 14, No. 3 (September 2005), pp. 1–17.

Raman, Varadaraja V. "Hinduism and Science: Some Reflections", *Zygon*, vol. 47, no. 3 (September 2012), pp. 549–574.

Raypa, R. S. "Bhotia" in K. Suresh Singh, Amir Hasan, B. R. Rizvi, and J. C. Das (eds.), *People of India. Uttar Pradesh.* New Delhi, India: Anthropological Survey of India, 2005, pp. 295–301.

Rodgers, Dennis. "Rhetoric versus Reality: 'Participatory Development', Cooperation and the Case of the Amul Cooperative", *Cambridge Anthropology*, Vol. 19, No. 1 (1996), pp. 73–91.

Sastry, S. S. "Central Himalayan Agriculture – A Case Study of Chamoli District" in S. B. Chakrabarti, B. R. Ghosh, and Ajit K. Danda (eds.), *Agrarian Situation in India: Volume Two.* Calcutta, India: Anthropological Survey of India, Ministry of Education and Culture, Govt. of India, 1984, pp. 71–76.

Sati, Vishwambhar P. and Kamlesh Kumar. *Uttaranchal: Dilemma of Plenties and Scarcities.* New Delhi, India: Mittal Publications, 2004.

Seth, Suman. "Putting Knowledge in Its Place: Science, Colonialism, and the Postcolonial," *Postcolonial Studies*, Vol. 12, No. 4 (2009), pp. 373–388.

Sharma, Sudha. "Communication Networking: An Initiative for Rural Children", *Vikas Vani: Voice for Sustainable Development*, Vol. 2, No. 4 (October–December 2008).

Shiva, Vandana. "Reductionist Science as Epistemological Violence" in Ashis Nandy (ed.), *Science, Hegemony and Violence: A Requiem for Modernity.* New Delhi, India: Oxford University Press, 1988.

Shivanna, K., K. Tirumalesh, J. Noble, T. B. Joseph, Gursharan Singh, A. P. Joshi, and V. S. Khati. "Isotope Techniques to Identify Recharge Areas of Springs for Rainwater Harvesting in the Mountainous Region of Gaucher Area, Chamoli District, Uttarakhand", *Current Science*, Vol. 94, No. 8 (April 2008).

Singh, Veer, Indira Ramesh, and Rachna Toshniwal. *Hamare Van, Pahad, aur Jeevan: Himalayavasi aur Jaiva-Vividhata.* Dehradun, India: Himalaya Trust, 2001.

Sinha, Subir. "Development Counter Narratives: Taking Social Movements Seriously" in K. Sivaramakrishnan and Arun Agrawal (eds.), *Regional Modernities: The Cultural Politics of Development in India.* Stanford, CA: Stanford University Press, 2003, pp. 286–312.

Souvenir – In Dedication to the Great Kar Sewa of Guru Nanak Dev Ji's Holy Kali Bein. Released by the President of India on August 17, 2006. Punjab, India: Ek Onkar Charitable Trust, Sultanpur Lodhi.

Turaga, Uday T. "STS: Indian Perspectives" in Carl Mitcham (ed.), *Encyclopedia of Science, Technology, and Ethics.* Detroit, MI: Macmillan Reference, 2005, pp. 986–999.

Veteto, James. "Anthropological and Philosophical Engagement in Overcoming the Nature/Culture Divide". University of North Texas Philosophy Colloquium, 4–26–13.

Weightman, Simon and S. M. Pandey. "The Semantic Fields of Dharma and Kartavy in Modern Hindi" in Wendy Doniger, J. Duncan, and M. Derrett (eds.), *The Concept of Duty in South Asia.* Columbia, MO: South Asia Books for the School of Oriental and Africa Studies, 1978.

Appendix 1

List of flora in the Himalayas

English Name	Hindi Name	Botanical Name
ringal (dwarf bamboo)	ringal	*Drepanostachyum falcatum*
agave	rambans	*Agave americana*
lantana	kuri, chamari	*Lantana camara*
amaranth	chaulai	*Amaranths gangeticus*
pine	chir	*Pinus roxburghii*
teak	saagon	*Tectona grandis*
buckwheat	kuttu	*Fagopyrum esculentum*
devadaru	devadar	*Cedrus deodara*
millet	mandwa	*Eleusine coracana*

Index